P9-CQY-854

Man and Nature in
the Renaissance

CAMBRIDGE HISTORY OF SCIENCE

Editors

GEORGE BASALLA
University of Delaware

WILLIAM COLEMAN
The University of Wisconsin

Man and Nature in the Renaissance

ALLEN G. DEBUS

Morris Fishbein Professor of the History of Science and Medicine

The University of Chicago

PUBLISHED BY THE PRESS SYNDICATE OF THE UNIVERSITY OF CAMBRIDGE
The Pitt Building, Trumpington Street, Cambridge, United Kingdom

CAMBRIDGE UNIVERSITY PRESS
The Edinburgh Building, Cambridge CB2 2RU, UK   http: //www.cup.cam.ac.uk
40 West 20th Street, New York, NY 10011-4211, USA   http: //www.cup.org
10 Stamford Road, Oakleigh, Melbourne 3166, Australia

© Cambridge University Press 1978

This book is in copyright. Subject to statutory exception and
to the provisions of relevant collective licensing agreements,
no reproduction of any part may take place without
the written permission of Cambridge University Press.

First published 1978
Reprinted 1980, 1981, 1983, 1985, 1986, 1987, 1988, 1990, 1991,
1992, 1993, 1994, 1995, 1996, 1999

Typeset in Garamond

A catalogue record for this book is available from the British Library

Library of Congress Cataloguing-in-Publication Data is available

ISBN 0-521-29328-6 paperback

Transferred to digital printing 2004

To my mother and father

Contents

Preface

No period in scientific history has been studied in greater detail than the Scientific Revolution, and yet it remains an enigma even as to its chronological limits. Some speak of a three-hundred-year time span stretching from 1500 to 1800 whereas others consider only the dramatic developments of the seventeenth century. The relation of the Renaissance to the Scientific Revolution is a crucial factor in any such dating, but in this volume we will speak of a scientific Renaissance dating roughly from the mid-fifteenth through the mid-seventeenth centuries. In this time span we will be able to follow the long-lasting and varied effects of humanism on medicine and the sciences and note also the continuing debate over a mystical view of nature espoused enthusiastically by alchemists and Hermeticists alike.

A work on Renaissance science might draw from many sources and it surely could reflect many viewpoints. It is frequently approached in terms of the progress of the exact sciences of mathematics and astronomy. Such studies in the past have generally paid little attention to the broader social and intellectual context of the period. Those authors who have emphasized the latter frequently downplay the importance of the technical scientific developments. In this volume our approach will be traditional in emphasizing the actual science of the period, but there will be frequent references to religion and to philosophical concepts that play little part in the science of the twentieth century. Thus we intend to discuss at some length the impact of alchemy and chemistry on the development of modern science and medicine, as these subjects have not yet been properly integrated into accounts of the scientific revolution. In fact the chemical debates of the early modern period generated more polemical literature than did those related to astronomy and the physics of motion. Because of this we must give

CONTROVERSIAL DEBATES (2 POLES)

ix

proper attention to these debates as well as to those that lead more directly to Galileo — and eventually, to Isaac Newton's *Principia mathematica*.

Certainly no attempt has been made here to present an exhaustive study of the period from 1450 to 1650. This is a volume in a series aimed at the student in Western civilization and our goal has been to produce a general survey, examining a number of key problems and themes. Our attention will be directed most frequently to the impact of humanism on the sciences, the search for a new method of science, and the continued dialogue between the proponents of a mystical-occult world view and those who sought a new kind of methematical-observational approach to nature.

The author is particularly grateful to the Newberry Library and the National Endowment for the Humanities for making it possible to complete this work in Chicago during the year 1975–1976 as the first step toward a more detailed treatment of the subject. The holdings of the Newberry Library are particularly valuable for the student of all aspects of Renaissance intellectual history, and William Towner, Richard H. Brown, and John Tedeschi have always been helpful in my quest for books, information, and such a variety of assistance that it would be hopeless to try to be more specific. The University of Chicago allowed me to take a leave of absence for the year — and, as always, I have received generous support from the Morris Fishbein Center for the Study of the History of Science and Medicine. Both editors of this series, George Basalla of the University of Delaware and William Coleman of the University of Wisconsin, have made helpful suggestions and the author acknowledges a special debt to William R. Shea of McGill University for the valuable comments he made on the first draft of this manuscript. In the final stages of preparation John Cornell and Russell H. Hvolbek prepared the index and read the text with care.

Deerfield, Illinois   *Allen G. Debus*
May 1978

CHAPTER I

Tradition and Reform

Few events in world history have been more momentous than the Scientific Revolution. The period between the mid-fifteenth and the end of the eighteenth centuries witnessed the growing cultural and political influence of Western Europe over all other parts of the globe. The new science and technology of the West was a crucial factor in this development, a fact recognized by most scholars at the time. Thus, Francis Bacon (1561–1626) observed in the *Novum organum* (1620) that

"it is well to observe the force and virtue and consequences of discoveries; and these are to be seen nowhere more conspicuously than in those three which were unknown to the ancients . . . ; namely, printing, gunpowder, and the magnet. For these three have changed the whole face and state of things throughout the world; the first in literature, the second in warfare, the third in navigation; whence have followed innumerable changes; insomuch that no empire, no sect, no star seems to have exerted greater power and influence in human affairs than these mechanical discoveries."

For Bacon these discoveries were Western in origin and relatively recent in date. He was neither the first nor the last to make such a statement, but there were few whose works were read more avidly by those who hoped to erect a new science in the seventeenth century.

But if the importance of the Scientific Revolution is readily admitted by all, the more we study its origins, the more unsure we become of its causes. In this volume we shall be concerned primarily with the two centuries from 1450 to 1650, the first date coinciding roughly with the beginning of the new humanistic interest in the classical scientific and medical texts and the second with the years just prior to the general acceptance of the mechanistic science of Descartes (1596–1650), Galileo (1564–1642), Borelli (1608–1679), Boyle (1627–1691), and Newton (1642–1727).

1

These two centuries present an almost bewildering maze of interests, and only rarely will an individual be found whose scientific methodology would prove to be fully acceptable to a modern scientist. Some of the scholars, whose work contributed to our modern scientific age, found magic, alchemy, and astrology no less stimulating than the new interest in mathematical abstraction, observation, and experiment. Today we find it easy — and necessary — to separate "science" from occult interests, but many then could not. And we cannot relegate this interest in a mystical world view to a few lesser figures forgotten today except by antiquarians. The writings of Isaac Newton and Johannes Kepler (1571–1630) reveal a genuine interest in transmutation and a search for universal harmonies no less than the work of Paracelsus (1493–1541), Robert Fludd (1574–1637), or John Dee (1527–1608). For the most part it has been traditional among historians of science to view their subject by hindsight, that is, to ignore those aspects of an earlier natural philosophy that no longer have a place in our scientific world. However, if we do this we cannot hope to reach any contextual understanding of the period. It will thus be our aim to treat this period in its own terms rather than ours. As we proceed we shall find that controversies over natural magic and the truth of the macrocosm — microcosm analogy were then as important as the better-remembered debates over the acceptance of the heliocentric system or the circulation of the blood.

Renaissance Science and Education

The very words "Renaissance" and "humanism" have been employed with so many connotations that there is little hope of satisfying any two scholars with a single definition. There is no need to try to do so here. To be sure, the Renaissance did involve a kind of "rebirth" of knowledge — no less than it did a rebirth of art and literature. And it was surely the period of the development of a new science. But having granted this, it is necessary to be careful to avoid simplification. The new love of nature expressed by Petrarch (d. c. 1374) and other fourteenth-century humanists had more than one effect. We readily accept that it was instrumental in the rise of a new observational study of natural phenomena, but we also find that Petrarch and later humanists deeply distrusted the traditional scholastic emphasis on philosophy and the sciences. The rhetoric and history they preferred was a conscious reply to the more technical "Aristotelian" studies that had long been the mainstay of the medieval university. The humanists sought the moral improvement of man rather than the logic and scholastic disputations characteristic of traditional higher learning.

These shifting values were to result in a new interest in educational problems. Fourteenth- and fifteenth-century reform programs were to be directed toward elementary education rather than the universities. The humanist educator Vittorino da Feltre (1378–1446) established a new school where students were urged to excel at sports and to learn military exercises. In classrooms they studied rhetoric, music, geography, and history — and, taking their examples from the ancients, they were taught to value both moral principles and political action above the basic principles of the trivium (grammar, rhetoric, and logic) or the study of traditional philosophical and scientific subjects.

Many of the most renowned humanist scholars were to be affected by this movement in educational reform. The result may be clearly seen in the work of Erasmus (1466–1536). He thought it enough for a student to learn of nature through his normal course of study in the reading of the ancient literary authors. Mathematics was not to him of much importance for an educated man. And Juan Luis Vives (1492–1540), surely the best known of all Renaissance educators, agreed fully when he argued against the study of mathematics that it tended to "withdraw the mind from practical concerns of life" and rendered it "less fit to fuse concrete and mundane realities."

But can we then say that the universities remained the centers of scientific training? For the most part they did, but there was an ever-increasing number of scholars both in medicine and the sciences who rejected the overwhelming conservatism of many — and perhaps most — of the institutions of higher learning. Peter Ramus (1515–1572) recalled his own academic training with despair:

"After having devoted three years and six months to scholastic philosophy, according to the rules of our university: after having read, discussed, and meditated on the various treatises of the *Organon* (for of all the books of Aristotle those especially which treated of dialectic were read and re-read during the course of three years); even after, I say, having put in all that time, reckoning up the years completely occupied by the study of the scholastic arts, I sought to learn to what end I could, as a consequence, apply the knowledge I had acquired wih so much toil and fatigue. I soon perceived that all this dialectic had not rendered me more learned in history and the knowledge of antiquity, nor more skillful in eloquence, nor a better poet, not wiser in anything. Ah, what a stupefaction, what a grief! How did I deplore the misfortune of my destiny, the barrenness of a mind that after so much labor could not gather or even perceive the fruits of that wisdom which was alleged to be found so abundantly in the dialectic of Aristotle!"

Ramus was not alone in his frustration — and his complaints were not without grounds. Paris, for example, was acknowledged as a stronghold of Galenic medicine in the sixteenth and seventeenth centuries whereas in England both the Elizabethan statutes for Cambridge (1570) and also the Laudian code for Oxford (1636) maintained the official authority of the ancients. Nor were the early professional societies necessarily better. The London College of Physicians looked on innovation with distrust. Thus, when in 1559 Dr. John Geynes dared to suggest that Galen (129/130–199/200 A.D.) might not be infallible, the reaction was immediate and severe. The good doctor was forced to sign a recantation before being received again into the company of his colleagues.

The conservatism seen in many major universities in the sixteenth and seventeenth centuries may be partially balanced by a critical tradition that had been applied to the ancient scientific texts at Oxford and Paris in the fourteenth century. This work, associated with scholasticism, was to prove particularly beneficial to the study of the physics of motion. As a scholarly tradition it was still in evidence at Padua and other northern Italian universities in the sixteenth century. For many, however, scientific criticism was a curious kind of humanistic game in which the scholar was to be commended for having eliminated the vulgar annotations and emendations of medieval origin that marred the texts of antiquity. His goal was textual purity rather than scientific truth.

In short, the educational climate in the early Renaissance was of questionable value for the development of the sciences. University training in this period may be characterized for the most part as conservative. As for the reform of primary education accomplished in the fourteenth and fifteenth centuries, this was openly antiscientific.

Humanism and Classical Literature

Dedication to the ancients is a familiar characteristic of Renaissance humanism. The search for new classical texts was intense in the fifteenth century, and each new discovery was hailed as a major achievement. No account is better known than that of Jacopo Angelo (fl. c. 1406). His ship sank as he was returning from a voyage to Constantinople made in search of manuscripts, but he managed to save his greatest discovery, a copy of the *Geography* of Ptolemy hitherto unknown in the West. Not long after this, in 1417, Poggio Bracciolini (1380–1459) discovered what was later to be recognized as the only copy of Lucretius's (c. 99–55 B.C.) *De rerum natura* to have survived from antiquity. This was to become a major stimulus for the revived interest in atomism two centuries later. And, just nine years after the recovery of Lucretius, Guarino da Verona (1370–1460) found a manuscript of the encyclopedic treatise on medicine by the second-century author, Celsus. This work, *De medicina,* was to exert a great influence, an influence due perhaps less to its medical content than to its language and style. This was the only major medical work to have survived from the best period of Latin prose and it was to be mined by medical humanists who sought proper Latin terminology and phrasing.

The search for new texts — and new translations — resulted in a new awareness of the importance of Greek. To be sure, Roger Bacon (c. 1214–1294) had already underscored this need in the thirteenth century, but the situation had not materially improved a century later. At that time Petrarch had lamented his own inadequate knowledge of this language. In fact he was not alone. Few Western scholars were able to use Greek until the teacher Manuel Chrysolorus (d. 1415) arrived in Italy with the Byzantine Emperor Manuel Paleologus in 1396. But helpful though Chrysolorus was, much greater enthusiasm was stirred by another Byzantine, Gemistos Plethon, on his arrival at the Council of Florence in 1439. The Greek revival was to affect all scholarly fields in the course of the fifteenth century. In medicine the humanist Thomas Linacre (c. 1460–1524) prepared Latin translations of Proclus (410–485) and of individual works of Galen. Significant though this was, his plans — only partially fulfilled — were actually far more grandiose. He projected a Latin translation of the complete works of Galen — and, with a group of scholars, a Latin translation of the complete works of Aristotle as well. Hardly less industrious was Johannes Guinter of Andernach (1505–1574), whose translations from Galen place him in the front rank of medical humanists. As professor of medicine at Paris, Guinter became one of the most prominent teachers of the young Andreas Vesalius (1514–1564).

This quest for truth in the search for accurate manuscripts was not confined solely to the study of the ancient physicians. Georg von Peuerbach (1423–1461) recognized the need for an accurate manuscript of Ptolemy's *Almagest* while writing his textbook, the *Theoricae novae planetarum.* But Peuerbach died while he was in the process of planning a journey to Italy to accomplish this end. His pupil, Johann Müller (Regiomontanus) (1436–1476) completed his master's journey and published an *Epitome* of the *Almagest.*

But Renaissance humanism cannot simply be reduced to the recovery of a pure Aristotle, Ptolemy, or Galen. No less influential on the development of modern science — and certainly part of the same humanistic movement — was the revival of the neo-Platonic, cabalistic, and Hermetic texts of late antiquity. So important did these seem to be that Cosimo de' Medici insisted that Marsilio Ficino (1433–1499) translate the recently discovered *Corpus hermeticum* (c. 1460) before turning to Plato or Plotinus. These mystical and religious works — to be discussed later in more detail — seemed to justify the pursuit of natural magic, a subject of great popularity among the savants of the sixteenth and seventeenth centuries. Included in this tradition was the call for a new investigation of nature through fresh observational evidence.

Coincidentally, this search for the pure and original texts of antiquity occurred when a new means existed for disseminating this knowledge, the printing press. It is interesting that the earliest printed book from Western Europe dates from 1447, at the very beginning of our period. For the first time it became possible to produce standard texts for scholars at a moderate price. In the scientific and medical fields these incunabula were for the most part printings of the old medieval scholastic texts scorned by the humanists. Thus the first version of Ptolemy's *Almagest* to be printed was the old medieval translation (1515). A new Latin translation appeared next (1528) — and finally the Greek text (1538), just five years prior to the *De revolutionibus orbium* of Copernicus. Galen and Aristotle were to proceed through the same stages.

The Growth of the Vernacular

Latin and Greek were surely the primary keys to the world of the scholar, but the Renaissance world was also characterized by a rapid growth in the use of the vernacular languages in learned fields. This is seen most strikingly in the religious pamphlets of the Reformation, where the author had an immediate need to reach his audience. But the use of the vernacular also became increasingly important in science and medicine in the course of the sixteenth century. This may be ascribed partially to the conscious nationalistic pride seen in this period. It is a time when authors wrote openly of their love of their native land and of their own language. A second factor was the feeling on the part of many of the need for a decisive break with the past. This seems to be ever more evident after the second quarter of the sixteenth century.

Recent research indicates a rapid increase in the use of the vernacular in the medical texts of the late Middle Ages. This trend intensified in the sixteenth century when a medical pamphlet war divided the Galenists from the Paracelsian medical chemists. This debate had been brought to the university level when Paracelsus lectured on medicine at Basel in his native Swiss-German in 1527. The medical establishment attacked him in force not only for the content of his lectures, but also for his choice of language. The latter was to remain a sore point among his followers for generations to come. Thus, the English Paracelsist Thomas Moffett (1553–1604) admitted — in Latin (1584) — that

"it is true that Paracelsus spoke often in German rather than Latin, but did not Hippocrates speak Greek? And why should they not both speak their native tongues? Is this worthy of reprehension in Paracelsus and to be passed over in Hippocrates, Galen and the other Greeks who spoke in their own language?"

The situation was not appreciably different in mathematics and the physical sciences. Galileo's publications in Italian remain classics of Italian literature today and in England numerous authors presented both popular and technical subjects in Tudor English. Of special interest is John Dee, who took it on himself to compose a preface to the first English translation of the *Elements of Geometry* by Euclid. Here he thought it necessary to explain that such a translation would pose no threat to the universities. Rather, he argued, many common folk might well for the first time be able "to finde out, and devise, new workes, straunge Engines, and Instrumentes: for sundry purposes in the Common Wealth or for private pleasure and for the better maintayninge of their owne estate." Similar apologies for the publication of scientific and medical texts in the vernacular are to be found in the other major modern languages from this period.

Observation and Experiment

Any general assessment of Renaissance science must include a discussion of a number of seeming paradoxes. A recurring theme in the sixteenth-century literature is the rejection of antiquity. But, as we have already noted, this rejection most commonly was directed at scholastic translations and commentaries. Some scholars did call for a completely new natural philosophy and medicine, but many adhered to the ancient philosophy — provided that they were assured that their texts were pure and unadulterated. There were those such as William Harvey (1578–1657), who openly praised the Aristotelian heritage. Others — and here Robert Fludd is a good example — attacked the ancients viciously while integrating many ancient concepts in their own work.

Also characteristic of the period was a growing reliance on observation and a gradual move toward our understanding of experiment as a carefully planned – and repeatable – test of theory. Older classics of observational science and method were recognized and praised by Renaissance scholars, who saw in them a model to be emulated. Thus many who rejected Aristotle's physics pointed to his work on animals as a text of major importance. Because of his use of observational evidence, Archimedes (287–212 B.C.) had great weight, whereas among medieval authors Roger Bacon, Peter Perigrinus (of Maricourt) (fl. c. 1270), and Witelo (Theodoric of Freiburg) (thirteenth century) were cited for their "experimental" studies.

Yet even though Roger Bacon and others might speak of a new use of observation as the basis for an understanding of the universe, it was far more customary to rely upon fabulous accounts related by Pliny the Elder (23–79 A.D.) or other ancient encyclopedists. Even the brilliant critique of the ancient physics of motion carried out at Oxford and Paris in the fourteenth century had been based more upon deductive reasoning and the rules of logic than upon the results of any new observational evidence.

The scientists of the sixteenth century did not immediately develop a modern understanding of the use of experiment, but there is evident in their work a more general recourse to observational evidence than existed before. Thus Bernardino Telesio (1509–1588) founded his own academy at Cosenza, which was dedicated to the study of natural philosophy. Rejecting Aristotle, whose work seemed to disagree with both Scripture and experience, he turned instead to the senses as a key to the study of nature. Of equal interest is John Dee, who numbered among his mathematical sciences *Archemastrie,* which "teacheth to bryng to actuall experience sensible, all worthy conclusions by all the Artes Mathematicall purposed. . . . And bycause it procedeth by *Experiences,* and searcheth forth the causes of Conclusions, them selues, in Experience, it is named of some *Scientia Experimentalis.* The *Experimentall Science."* Here the word "experimental" may best be understood as "observational." The concept of the modern controlled experiment was not part of Dee's methodology.

Mathematics and Natural Phenomena

Surely no less important than the new appreciation of observational evidence was the development of quantification and the increasing reliance on mathematics as a tool. Plato had stressed the importance of mathematics, and the revived interest in his work did influence the sciences in this area. In our period Galileo stands as the key figure in this development. Viewing mathematics as the essential guide for the interpretation of nature, he sought a new description of motion through the use of mathematical abstraction. In doing so Galileo was acutely aware that he was departing from the traditional Aristotelian search for causes.

Combined with the novel use of mathematics in natural philosophy, there were dramatic new developments within mathematics itself. The work of Tartaglia (1500–1557), Cardano (1501–1576), and Viète (1540–1603) in algebra did much to advance that subject in the sixteenth century — and tedious arithmetical calculations were greatly simplified through the invention of logarithms by Napier (1550–1617). And only slightly beyond our period comes the invention of the calculus by the independent efforts of Leibniz (1646–1716) and Newton. All these tools were quickly seized upon by contemporary scientists as aids to their work.

If one were to ask the reasons for this use of mathematics in the sixteenth century, one might arrive at a variety of answers. One would surely be the new availability of the work of Archimedes, the Greek author whose approach most closely approximated that of the new science. His texts had never been completely lost, but there is clear evidence of a new Archimedean influence in the mid-sixteenth century with a series of new editions of his work. Another factor of importance is the persistence of interest in the study of motion initiated by the fourteenth-century scholars at Oxford and Paris. There seems little doubt that Galileo was as a student the beneficiary of this tradition. A third factor was surely the Platonic, neo-Platonic, and Pythagorean revival. This influence often had a mystical flavor, but whatever its form, it was an important stimulus for many scientists of the period. And finally, one might point to the need for practical mathematics associated with the practical arts and technology.

Technology

It is rewarding to pause momentarily to examine this new interest in technology. While the extent of the relationship is open to debate, it is clear that at the very least those interested in warfare required mathematical studies in their use of cannon, and the navigator had to perform calculations to determine his position at sea. This was a period that witnessed impressive advances in instrumentation, ranging from practical astrolabes for the mariner to the massive astronomical instruments built by Tycho Brahe. The telescope, the microscope, the first effective thermometers, and a host of other tools were developed by artisans and scientists alike. Indeed, the scientists were taking an active interest in the work of the tradesmen for the first time. This may be interpreted partially as a revolt against the authority of the ancients, as most ancient and medieval studies of nature were totally divorced from processes employed by workmen. The scholastic student of the medieval university agreed with the ancients and rarely left his libraries and study halls. In the Renaissance, however, we witness a great change. There may be few descriptions of the practical arts in the books of the fifteenth century, but handbooks of mining operations began to appear from the presses as early as 1510 and similar works relating to other fields appeared shortly thereafter.

In contrast to earlier periods the scientists and physicians now acknowledged openly that the scholar would do well to learn from the common man. Paracelsus advised his readers that

"not all things the physician must know are taught in the academies. Now and then he must turn to old women, to Tartars who are called gypsies, to itinerant magicians, to elderly country folk and many others who are frequently held in contempt. From them he will gather his knowledge since these people have more understanding of such things than all the high colleges."

And Galileo candidly began his epoch-making *Discourses and Demonstrations Concerning Two New Sciences* (1638) with the following statement:

"The constant activity which you Venetians display in your famous arsenal suggests to the studious mind a large field for investigation, especially that part of the work which involves mechanics; for in this department all types of instruments and machines are constantly being constructed by many artisans, among whom there must be some who, partly by inherited experience and partly by their own observations, have become highly expert and clever in explanation."

This list could be greatly amplified if we took into account the great mining treatises of Agricola (1494–1555) and Biringuccio (fl. c. 1540), the views of Francis Bacon on the practical purpose of science, and the stated practical goals of the early scientific societies. There is little doubt that some areas of science progressed because the contribution of artisans and scientists fostered the study of practical processes. Johann Rudolph Glauber (1604–1670) was so encouraged by the developments he had witnessed that he forecast the supremacy of Germany over all Western Europe if its rulers would only follow his plan outlined in the *Prosperity of Germany*. And yet, even if we grant this belated recognition of technology by the scientist, there was no appreciable feedback from the small scientific community to technology until well into the eighteenth century.

Mysticism and Science

A fourth ingredient in the formation of the new science — and a most un-likely one from our post-Newtonian vantage point — was the new Renais-sance interest in a mystical approach to nature. Much of this may be at-tributed to the strong revival of interest in the Platonic, neo-Platonic, and Hermetic writings. It is instructive to note this influence first in mathe-matics and then in the widespread interest in natural magic.

From our point of view Renaissance mathematics had the effect of a double-edged sword. On the one hand, the new interest in mathematics furthered the development of a mathematical approach to nature and the internal development of geometry and algebra; on the other hand, the same interest resulted in occultist investigations of all kinds related to number mysticism. Renaissance cabbalistic studies encouraged a mystical numero-logical investigation of the Scriptures with the hope that far-reaching truths would be found. Similarly magic squares and harmonic ratios seemed to offer insight into nature and divinity. Even in antiquity this ten-dency was embodied in the Pythagorean tradition prior to the time of Plato. The latter's numerological speculations in the *Timaeus* were to con-tinue to affect the world of learning throughout the Middle Ages, and with the revival of the texts of late antiquity in the fifteenth century the same themes were heard once again.

It is important not to try to separate the "mystical" and the "scientific" when they are both present in the work of a single author. To do so would be to distort the intellectual climate of the period. Of course it is not dif-ficult to point to the mathematical laws governing planetary motions for-mulated by Kepler or the mathematical description of motion presented by Galileo. These were basic milestones in the development of modern science. But it should not be forgotten that Kepler sought to fit the orbits of the planets within a scheme based upon the regular solids, and Galileo was never to relax his adherence to circular motion for the planets. Both au-thors reached conclusions that were strongly influenced by their belief in the perfection of the heavens. Today we would call the first examples "sci-entific," the second not. But to force our distinction upon the seventeenth century is ahistorical.

Robert Fludd offers an excellent example of an Hermetic–chemical ap-proach to mathematics. Few would have insisted more than he that mathe-matics was essential for any study of the universe. But Fludd would have added that the true mathematician should lift his sights high. His aim should be to show the divine harmonies of nature through the interrelationship of circles, triangles, squares, and other figures. These would clearly indicate the connections of the great world to man. Fludd sought a new approach to nature and, like Kepler and Galileo, he hoped to use mathematics as a key, but for him quantification was something quite different than it was for the others. Fludd believed that the mathematician should use this tool to study the overall design of the universe. He should not — like Galileo — be concerned with lesser phenomena such as the motion of a falling object.

The case of mathematics is of special importance because of the significance of quantification in the rise of modern science, but the occult or mystical influence of late Hellenistic philosophy had a much deeper impact on sixteenth-century thought than this alone Implicit in neo-Platonism and the Christian traditions was the belief in a unity of nature, a unity that encompassed God and the angels at the one extreme and man and the terrestrial world at the other. Along with this was a continued belief in the truth of the macrocosm—microcosm relationship, the belief that man was created in the image of the great world, and that real correspondences do exist between man and the macrocosm.

The general acceptance of the macrocosm and the microcosm along with the great chain of being gave credence to the acceptance of correspondences existing everywhere between the celestial and sublunary worlds. In the ancient world such beliefs seemed to give a solid basis for astrology. It seemed reasonable to assume that stars should influence mankind here on earth. In the Renaissance many agreed, astral influences did indeed affect the earth and man. The Hermetic texts added a new ingredient to this world view. Largely on their basis man was now viewed as a favored link in the great chain of being. Partaking in Divine Grace, he was something more than the passive recipient of starry influences. And, as there is a general sympathy between all parts of the universe, man may affect supernature as well as be affected by it. This concept had immediate value in medicine through the doctrine of signatures. Here it was postulated that the true physician had the power successfully to seek out in the plant and mineral kingdoms those substances that correspond with the celestial bodies, and therefore ultimately with the Creator.

All this is closely related to the basis of Renaissance natural magic. The true physician of the Paracelsus or Ficino type was at the same time a magician who conceived nature to be a vital or magic force. Such a student of nature might learn to acquire natural powers not known to others and thus astonish the populace, even though these powers were known to be

God-given and available to all. Indeed, to many this seemed to be one of the most attractive aspects of magic. Thus, late in life, John Dee recalled his student days at Cambridge where he had prepared a mechanical flying scarab for a Trinity College performance of Aristophanes' *Peace*, "whereat was great wondering, and many vaine reportes spread abroad of the meanes how that was effected." Dee's scarab was in the tradition of Hellenistic mechanical marvels, but he was also well aware that true magic meant the observational study of the unexplained or occult forces of nature. Thus in his *Natural Magick* John Baptista Porta (1540–1615) had explained that magic is essentially the search for wisdom and that it seeks nothing else but the "survey of the whole course of nature." Still earlier Heinrich Cornelius Agrippa (c. 1486–1535) called this the most perfect knowledge of all, and Paracelsus equated it with nature itself and spoke of it in terms of a religious quest that would lead the seeker to a greater knowledge of his Creator.

For such men natural magic was far removed from the taint of necromancy. Rather, magic was closely associated with religion through the search for divine truths in created nature. Nevertheless, the scientist who was willing to accept the title of "magician" might well expose himself to danger. Again John Dee will serve as an example. Imprisoned early in life for his active interest in astrology, he was later to have his vast library destroyed by an angry mob. Appealing to his readers for sympathy, he asked whether they could really think him such a fool as "to forsake the light of heauenly Wisedome: and to lurke in the dungeon of the Prince of darknesse?" Despite the accusations that had been made, he held himself to be "innocent, in hand and hart: for trespacing either against the lawe of God, or Man, in any of my Studies of Exercises, Philosophicall, or Mathematicall."

In reality sixteenth-century natural magic was a new attempt to unify nature and religion. For the Hermeticists and the natural magicians the works of Aristotle were flawed by heretical concepts, and they were repeatedly to recall that church councils had condemned many of these Aristotelian errors. This being the case, why should Aristotle and Galen still be the basis of university teaching when there was another interpretation of nature through natural magic and occult philosophy — subjects whose very existence depended upon the sacred Scriptures? How could it be that any Christian should prefer the atheistic Aristotle to this new and pious doctrine? In truth, they argued, knowledge may be acquired by Divine Grace alone; either by some experience such as St. Augustine's divine illumination or else by means of experiment in which the adept might attain his end with the aid of divine revelation. The religious content of early-seventeenth-century Hermeticism is evident in the work of Thomas Tymme (d. 1620), who wrote (1612) that

"the Almighty Creator of the Heavens and the Earth . . . hath set before our eyes two most principal books: the one of nature, the other of his written Word . . . The wisdom of Natures book, men commonly call Natural Philosophy which serveth to allure to the contemplation of that great incomprehensible God, that wee might glorify him in the greatness of his work. For the ruled motions of the Orbes . . . the connection, agreement, force, virtue, and beauty of the Elements . . . are so many sundry natures and creatures in the world, are so many interpreters to teach us, that God is the efficient cause of them and that he is manifested in them, and by them, as their final cause to whom also they tend."

This was written to explain why he had prepared a book devoted to nature, the generation of the elements, and other essentially scientific topics. For an author such as Tymme science and the observation of nature were a form of divine service, a true link with divinity. In a sense natural research was a quest for God.

The student of Renaissance science must thus cope with more than the work of Copernicus and its consequences or the anatomical research leading to the discovery of the circulation of the blood. As for scientific method, the historian must concern himself with the new interest in mathematics and quantification all the while taking care not to divorce it from subjects as alien to modern science as the doctrine of signatures and natural magic. Indeed, our science today owes much to that search for a new synthesis of man, nature, and religion, which characterized the work of many scientists and physicians four centuries ago.

Renaissance science and medicine were deeply influenced by three figures of the sixteenth century and three others from antiquity. The first three were Nicholas Copernicus (1473–1543), Andreas Vesalius, and Philliptus Aureolus Theophrastus Bombastus von Hohenheim, called Paracelsus — the last three Archimedes, Galen, and Ptolemy. All were to register their impact on the learned world at approximately the same time. Indeed, the *De revolutionibus orbium* (Copernicus), *De humani corporis fabrica* (Vesalius), and the first major translation into Latin of the works of Archimedes all appeared in 1543.

The work of Paracelsus began to affect the learned world shortly after his death in 1541 when his scattered manuscripts were collected and published extensively for the first time. It is to his work that we turn next since to a greater extent than the others, Paracelsus may be viewed as a herald of the Scientific Revolution. And yet, though his call for a new approach to nature was coupled with a venomous attack on the followers of the ancients, Paracelsus himself was typical of the Renaissance in his willingness to borrow freely from the very texts and authors he rejected in print.

IATROCHEMISTRY: A SCHOOL OF THOUGHT OF THE 16TH & 17TH THAT SOUGHT TO UNDERSTAND MEDICINE & PHYSIOLOGY IN TERMS OF CHEMISTRY.

POLEMICS- THE ART OR PRACTICE OF ENGAGING IN CONTROVERSIAL DEBATE OR DISPUTE

THREE PRINCIPLES

CHAPTER II

The Chemical Key

A new interest in chemistry is very noticeable in the late Renaissance. Relatively few chemical books had been published in the period prior to 1550, but during the next century a veritable flood of chemical and medico—chemical texts was printed. Those who wrote these books or who printed older texts insisted on the importance of their work. They not only spoke of the large number of those who had abandoned the teachings of the ancients to follow their chemical philosophy, but they also frequently named the chemical authorities to whom their readers could turn for truth in philosophy and medicine. They all hoped that the doctrines of the ancients would soon be overturned and that their "new philosophy" of nature would triumph. On the other side, scientists as prominent as Johannes Kepler and the early mechanists Marin Mersenne and Pierre Gassendi were to write at length against the mystical philosophy of nature elaborated by the chemists. But why was chemistry such a center of debate? The immediate answer may be found in the controversial writings of Paracelsus, but to understand him we must look briefly at the chemical background to his work.

The Chemistry of the Latin West

Chemical texts were introduced to Western Europe in the twelfth century along with other treasures of Greek science, philosophy, and medicine by way of translation or abstraction (for the most part) from the Arabic. The early translations already characterize chemistry as a secret art, so secret that it is often difficult if not impossible to identify the original texts from which they were made. But when we pass beyond the indistinct scene of the twelfth century, we become aware of a rapidly increasing interest in this subject throughout the next two centuries prior to a decline in the quantity — and quality — of new texts in the fifteenth century. There are numerous references to alchemical allegory in medieval literature, and Chaucer's late-fourteenth-century "Canon Yeoman's Tale" remains the best description of the gold-making charlatan.

Medieval alchemy incorporated much Aristotelian doctrine. The four elements of earth, water, air, and fire had served not only as the basis of Aristotelian physics but, in the form of the related four humors (blood, phlegm, yellow, and black bile), had also served as the bedrock of Galenic medical theory. The qualities associated with these elements (hot, cold, wet, and dry) were interchangeable and thus permitted the transmutation of one element into another. Islamic scholars had added to this a new theory of metals in the eighth century. They had taught that metals were composed of a hypothetical philosopher's (not actual) mercury and sulfur. When these two occurred in perfect proportion the resultant metal would be gold.

But along with Aristotelian and Islamic element theory, alchemy carried with it an aura of secrecy and mysticism. This may in part be a carry-over of the atmosphere in which the early metalworkers of Egypt operated. But a second source might well be the occult tradition of the mystical religions of late antiquity. Gnostic, neo-Platonic, and neo-Pythagorean elements did much to differentiate these alchemists from the students of optics, astronomy, and mathematics. And surely the religious currents run deep in the alchemical literature. The great work itself was seen as a religious experience, and processes and substances were frequently explained in terms of soul, body, and spirit.

Along with this allegorizing and mysticism the alchemist placed a new emphasis on observational evidence. We have already noted Paracelsus's call for the adept to learn from nature rather than books, but the same message is evident in the earlier literature. The fourteenth-century alchemist Bonus of Ferrara observed that

"If you wish to know that pepper is hot and that vinegar is cooling, that colocynth and absinthe are bitter, that honey is sweet, and that aconite is poison; that the magnet attracts steel, that arsenic whitens brass, and that tutia turns it of an orange color, you will, in every one of these cases, have to verify the assertion by experience. It is the same in Geometry, Astronomy, Music, Perspective, and other sciences with a practical aim and scope. A like rule applies with double force in alchemy, which undertakes to transmute the base metals into gold and silver . . . The truth and justice of this claim, like all other propositions of a practical nature, has to be demonstrated by a practical experiment, and in no other way can it satisfactorily be shown."

Related to this emphasis on observation was the alchemist's interest in laboratory procedures. The medieval period saw great advances in distillation techniques. Far more efficient furnaces were built than had been available earlier. Now, with higher temperatures and better condensation, it was possible to add new reagents (most notably alcohol and the mineral acids) to the chemical laboratory. The Latin Geber (early fourteenth century, an assumed name referring to the eighth-century Jābir ibn Hayyān) produced the outstanding work of this sort describing equipment and chemical processes.

Geber made little reference to medicine, but this connection was to become an important aspect of medieval alchemy. The search for chemicals of pharmaceutical value appears among Islamic authors in the writings of the al-Rāzī (Rhazes) (c.854–925/926) and then frequently among his followers. In the West Roger Bacon noted in the *Opus tertium* (1267) that although many physicians used chemical processes to prepare their medicines, very few knew how to perform those works which led to the prolongation of life. His younger contemporary Arnald of Villanova (c.1235–1311) and the fourteenth-century author John of Rupescissa continued to emphasize the medical value of chemistry. By the early sixteenth century this form of scientific literature had blossomed into the many distillation books so very characteristic of the period. These all included descriptions of the chemical equipment needed for the production of oils and spirits from plant substances of all kinds. The benefits of these "quintessences" seemed so great that sixteenth-century editions of the ancient herbal of Dioscorides were furnished with a chemical appendix to bring it up to date.

This chemical knowledge was not considered to be in any way opposed to the science of the Aristotelians or the medicine of the Galenists. To be sure, there were some who complained about the conservatism of the schools, but alchemy had come to the West along with the rest of the body of ancient knowledge. It had been cultivated with classical philosophy and medicine in the Near East — and it was not to be immediately divorced from this earlier union. Nor is there any indication that chemistry was viewed as a dangerous rival discipline by physicians or natural philosophers.

The translation of the *Corpus hermeticum* by Ficino in 1463 added one more factor that was to affect the chemistry of the Renaissance. Fostering occult learning of all kinds, alchemy was shortly brought to the attention of all learned men as an area of study that had not received proper attention in the past. Both Heinrich Cornelius Agrippa von Nettesheim and John Baptista Porta were to point to alchemy as a fundamental science for the understanding of nature. John Dee applied "geometrical method" through twenty-four theorems to the construction of his "hieroglyphic monad," a figure that closely approximated the alchemical symbol for mercury. In the course of this construction Dee felt that he had repeated the first steps of the Creation. The reader was promised the understanding of great mysteries and the entire work would appear to be no less than a veiled representation of the alchemical process itself. But Dee's emphasis was clearly of a piece with the spiritual mathematics favored by those Renaissance Pythagoreans who sought a key to the Creation in mysticism and numerical analysis. The truths of magic were avowed whereas more conventional mathematical proofs, chemical laboratory techniques, and practical medical applications were of relatively little interest. It was in this way that Dee felt that alchemy might be recognized as the most fundamental subject for the natural philosopher.

Nearly half a century earlier Paracelsus had found in alchemy a new basis for the theory of medicine. This, in turn, was to be developed into a universal philosophy of nature validated by the natural correspondences linking man and the world about him. And if Dee's mystical, "mathematicized" alchemy had little impact beyond a devoted circle of alchemists, the views of Paracelsus were to lead to a European debate relating to both medicine and natural philosophy.

Paracelsus: A Lifetime Search

Born near Zurich in the small town of Einsiedeln in 1493, Philippus Aureolus Theophrastus Bombastus von Hohenheim was only later referred to as "Paracelsus," or "greater than Celsus." As a child he was exposed to a heady mixture of Renaissance thought. His father was a country physician who dabbled in alchemy, and the son was never to lose his interest in either medicine or the chemical laboratory. Young Paracelsus was to study under the famed abbot and alchemist, Johannes Trithemius (1462–1516), and was to learn the lore of mines when he worked as an apprentice in the Fugger mines in Villach when his father moved there in 1500. This experience was to bear fruit later in his speculations on the growth of metals and his book on the diseases of miners, the first book ever written on an occupational health problem.

At the age of fourteen Paracelsus left home to study, and over a period of more than two decades he traveled widely. He visited many universities and may have received a medical degree at Ferrara, but if so, he was willing to work in the far less prestigious position of surgeon with the armies that were ever on the move throughout Europe. By the third decade of the century his travels become easier to follow. Now in his thirties, he confined his travels to Central Europe, where he moved constantly from town to town both writing and offering his services as a physician. There were occasional moments of glory such as his appointment as municipal physician at Basel in 1527, but these were always short due to his rash temper. He made no effort to disguise his contempt for the universities and their academic circles. As for the physicians, they need hardly be considered:

"I need not don a coat of mail or a buckler against you, for you are not learned or experienced enough to refute even one word of mine . . . you defend your kingdom with belly-crawling and flattery. How long do you think this will last? . . . Let me tell you this: every little hair on my neck knows more than you and all your scribes, and my shoe-buckles are more learned than your Galen and Avicenna, and my beard has more experience than all your high colleges."

Such outbursts were to lose him position after position, for they offended even those who most wanted to help him. As a result he was constantly on the move; he died in Salzburg in 1541, where he had only recently been called by the bishop suffragan Ernest of Wittelsbach.

The Paracelsian Chemical Philosophy

At the death of Paracelsus there was little to indicate that his work would become the focal point for debate among scholars for more than a century. True, he had been a controversial figure during his lifetime, but relatively few of his voluminous writings had been published while he had been alive. The flood of Paracelsian texts began to issue from the presses only later. The legend of the man's near-miraculous cures began in the years after 1550 and soon there was a widespread search for his manuscripts, which were often published with notes and commentaries. Toward the end of the century vast collected editions were printed, and a whole school of Paracelsians battled with Aristotelians and Galenists over the course of natural philosophy and medicine alike.

Because of the late publication of the texts it is as proper to speak of the philosophy of the Paracelsians as it is that of Paracelsus. But even if we make this allowance the chemical philosophy is difficult to reconstruct, partly because no simple textbooks were published and partly because the views of these men are alien to those of the twentieth-century scientist.

Actually there is much in the work of the Paracelsians that is reminiscent of other Renaissance natural philosophers. Above all, they sought to overturn the traditional, dominant Aristotelianism of the universities. For them Aristotle was a heathen author whose philosophy and system of nature was inconsistent with Christianity, a point of considerable concern during the Reformation. They stated that his influence on medicine had been catastrophic because Galen had uncritically accepted his work and the Aristotelian–Galenic system had subsequently become the basis of medical training throughout Europe. For them the universities were hopelessly moribund and unyielding in their adherence to antiquity.

The Paracelsians hoped to replace all this with a Christian neo-Platonic and Hermetic philosophy, one that would account for all natural phenomena. They argued that the true physician might find truth in the two divine books: the book of divine revelation – Scripture – and the book of divine Creation – nature (Figure 2.1). Thus, the Paracelsians applied themselves on the one hand to a form of biblical exegesis, and on the other to the call for a new philosophy of nature based on fresh observation and experiment. An excellent example of this may be found in the work of the important early systematizer of the Paracelsian corpus, Peter Severinus (1540–1602), physician to the king of Denmark, who told his readers that they must sell their possessions, burn their books, and begin to travel so that they might make and collect observations on plants, animals, and minerals. After their *Wanderjahren* they must "purchase coal, build furnaces, watch and operate with the fire without wearying. In this way and no other you will arrive at a knowledge of things and their properties."

One senses a strong reliance on observation and experiment in the work of these men even though their concept of what an experiment is and its purpose was often quite different from our own. At the same time one notes an underlying distrust of the use of mathematics in the study of nature. They might well, as Platonists, speak of the divine mathematical harmonies of the universe. Paracelsus, in addition, spoke firmly of true mathematics as the true natural magic. But it was more customary for the Paracelsians to react with distaste to the logical, "geometrical," method of argument employed by the Aristotelians and Galenists. They condemned this "mathematical method" along with the traditional scholastic emphasis on geometry and they very specifically attacked mathematical abstraction in the study of natural phenomena — particularly the study of local motion. Their reason for this was primarily religious, and they were particularly in-

Figure 2.1. The true chemical philosopher learns through divine revelation as well as chemical studies. From Heinrich Khunrath, *Amphitheatrum sapientiae* (1609). From the collection of the author.

censed by the *Physics* of Aristotle. There – through the study of motion – it was argued that the Creator God must be immobile. The Paracelsian chemists of the period of the Reformation stated firmly that any argument imposing such a restriction on the omnipotent Deity could not be accepted — and that for this reason alone the texts of the ancients were sacrilegious and must be discarded. The chemical philosophy was to be a new science based firmly on observation and religion. Those who turned to quantification might recall that God had created "all things in number, weight and measure." This was interpreted as a mandate for the physician, the chemist, and the pharmacist — men who weighed and measured regularly in the course of their work (Figure 2.2).

If the Paracelsians rejected what they called the "logico–mathematical" method of the schools, they turned to chemistry with the conviction that this science was the basis for a new understanding of nature. It was an observational science, and its scope was universal. These claims were to be found in the traditional chemical texts. For Paracelsus alchemy had offered an "adequate explanation of all the four elements," and this meant literally that alchemy and chemistry might be used as keys to the cosmos either through direct experiment or through analogy. Paracelsus explained the Creation itself as a chemical unfolding of nature. The later Paracelsians agreed and amplified this theme. Gerhard Dorn (fl. 1565–1585) gave a detailed description of the first two chapters of Genesis in terms of the new chemical physics, and Thomas Tymme argued that the Creation had been nothing but an "Halchymicall Extraction, Separation, Sublimation, and Conjunction."

The chemical interpretation of Genesis helped to focus attention on the problem of the elements as the required first fruit of the Creation. Although the Paracelsian *tria prima* (salt, sulfur, and mercury) was a modification of both the earlier sulfur-mercury theory of the metals and other elemental triads, it has a special significance in the rise of modern science. The Aristotelian elements (earth, water, air, and fire) served as the basis of the accepted cosmological system. They were used by the alchemists as a means of explaining the composition of matter, by the physicians (through the humors) as a system for the interpretation of disease, and by the physicists as the basis for the proper understanding of natural motion. The introduction of a new elemental system thus ran the risk of calling into question the whole framework of ancient medicine and natural philosophy.

Although the new principles can properly be interpreted as part of an attack on scholastic philosophy, it is clear also that they led to considerable confusion. Paracelsus had not clearly defined these principles, and, indeed, they were of little value in the development of modern analytical chemistry inasmuch as they were described as differing qualitatively in different materials. Nor had Paracelsus offered the principles specifically as a replacement for the Aristotelian elements. Rather, he had used both systems —

Figure 2.2. The earliest illustration of an enclosed analytical balance is to be found in this picture of an alchemical laboratory. From the *Theatrum Chemicum Britannicum*, ed. Elias Ashmole (1652). Courtesy of the Department of Special Collections, The University of Chicago.

and often in a seemingly contradictory fashion. By the fourth quarter of the sixteenth century we find element theory in a state of flux, with chemists choosing from observational evidence and Paracelsian texts as they saw fit. Nevertheless, from the texts of this period we can see that the chemical physicians were turning in increasing numbers to the three principles as a means of explanation. Some were attracted by the trinitarian analogy of body, soul, and spirit, whereas others turned to them in search of an alternative to the humors. For chemical theorists they represented philosophical substances that might never be isolated in reality, whereas for the practical pharmacist they were nothing else but his distillation products. It was not uncommon for a medicinal herb to yield a watery phlegm, an inflammable oil, and a solid, and it was felt that these at least indicated the presence of the primal principles of mercury, sulfur, and salt.

The concept of a chemical universe went beyond the chemical interpretation of the Creation and the problems of element theory. Those authors interested in meteorology explained thunder and lightning as a combination of an aerial sulfur and niter analogous to the explosion of sulfur and saltpeter in gunpowder. Similarly, Paracelsian authors were the first to offer a hypothesis meaningful for the development of agricultural chemistry. Seeking a cause for the beneficial effects of manuring in farming, they correctly postulated that the manure offered essential soluble salts to the soil.

Indeed, for the Paracelsians, the earth was seen as a vast chemical laboratory and this explained the origin of volcanoes, hot springs, mountain springs, and the growth of metals. The old concept of an internal fire was given as the explanation of volcanoes, which were understood as the eruptions of molten matter through surface cracks (Figure 2.3). Mountain streams were explained in an analogous fashion. Here they argued that subterranean water reservoirs were distilled by the heat of the central fire. As this vapor reached the surface, mountains acted as chemical alembics, and the result was the "distilled" mountain stream. Yet some rejected the possibility of such a fire, arguing that the air requisite for such a conflagration did not exist within the earth. Henri de Rochas (fl. 1620–1640) suggested that the heat of mineral-water springs derives from the reaction of sulfur and a nitrous salt in the earth. The English physician Edward Jorden (1569–1632) offered a more comprehensive chemical alternative. A thorough vitalist like most chemists of the period, Jorden accepted the commonly held notion of the growth of metals, but accounted for it in a new way. He turned to the alchemical process of "fermentation," which he defined as a heat-producing reaction requiring no air. This, he argued, must be the cause of inorganic growth. This new source of heat enabled one

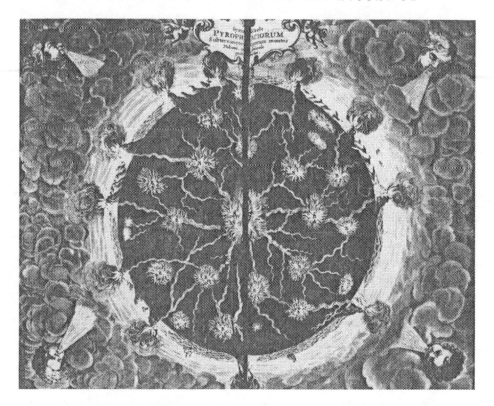

Figure 2.3. Diagram depicting the interrelation of volcanoes and the central fire. From Athanasius Kircher, *Mundus subterraneus* (1678). Courtesy of the Department of Special Collections, The University of Chicago.

to understand volcanoes and mountain streams without the troublesome notion of the central fire.

The Microcosm and Medical Theory

The Paracelsian chemical philosophy was considered to be a new observational approach to all nature, but from the beginning it had a special appeal for physicians. Paracelsus had insisted that God rather than the constellations had created him a physician; his followers repeated this and added that because of its divine origin medicine stood above the other sciences. Here they both reflected the priest–physician concept of Renaissance neo-

Platonism, and it is likely that their ultimate source may be found in Ecclesiasticus 38:1, "Honor the physician for the need thou hast of him: for the Most High hath created him." Indeed, for Paracelsus the role of the physician might properly be compared with that of the true natural magician.

Paracelsus and his early followers firmly believed in the macrocosm-microcosm analogy. Man is a small replica of the great world about him, and within him are represented all parts of the universe (Figure 2.4). At all times it was considered fruitful to seek out correspondences between the greater and lesser worlds, and the theory of sympathy and antipathy was employed to explain universal interaction. In contrast to Aristotelians, who insisted on action through contact, the Paracelsians found no difficulty in accepting action at a distance. It is thus easy to understand why Paracelsian Hermeticists should have been among the first to defend the experimental research of William Gilbert on the magnet. In the field of medicine, the controversial weapon salve cure (cure by sympathy involving treatment of the weapon rather than the wounded person) surely assumed the possibility of action at a distance.

For the Paracelsian the humoral theory of Galenic medicine was no longer adequate. The traditional explanation of disease as an internal imbalance of the humors was rejected by Paracelsus. He preferred to emphasize local malfunctions within the body that were ascribed to one of the three principles. A major cause of disease for him was to be found in external seedlike factors that were introduced to the body through the air, food, or drink. These localized themselves and then grew in specific organs. Here an analogy could be drawn between the macrocosm and the microcosm. In the same fashion that metallic "seeds" in the earth resulted in the growth of metallic veins, "seeds" of disease grew within the body while they combated the local life force of a specific organ. This life force separated pure substance from waste in a manner analogous to the alchemist who sought to isolate pure quintessences from gross matter in his laboratory.

The relationship of the macrocosm to man had further chemical implications. The French Paracelsian Joseph Duchesne (c. 1544–1609) exemplified the persistent search for chemical analogies among the Paracelsians when he spoke of respiratory diseases in terms of the same distillation analogy utilized by other iatrochemists (or medical chemists) when explaining the origin of mountain streams. Special significance was attached to the air, recognized as essential for the maintenance of both fire and life. If, on the one hand, an aerial sulfur and niter might combine to cause thunder and lightning in the sky or hot springs in the earth, on the other hand, they might react within the body when inhaled to generate diseases charac-

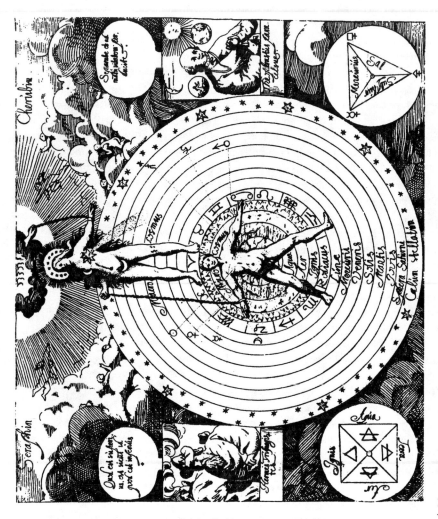

Figure 2.4. Man as the microcosm joined to his creator by the chains of nature, depicted as a young woman. Note the portraits of Hermes and Paracelsus plus the diagrams of the four elements and the three principles. From Tobias Schütz, *Harmonia macrocosmi cum microcosmi* (1654). Courtesy of the Department of Special Collections, The University of Chicago.

terized by hot and burning qualities (Figure 2.5). By the early seventeenth century aerial niter had become associated with a life force requisite for man. Indeed, this life force was on occasion identified with the *spiritus mundi.* It was postulated that after having been separated from gross air in the lungs, this substance was formed into arterial blood. Inasmuch as they maintained this concept — or modifications of it — it is little wonder that we find seventeenth-century Paracelsians rejecting the common practice of bloodletting. This operation, they argued, would only diminish the essential life force of the patient. At the same time, the rejection of bloodletting served to reflect their opposition to traditional humoral pathology.

If the Paracelsian chemical philosophy of nature provided a conceptual framework for the iatrochemist, it also provided a basis for his practical work. Because of the importance of heat and of fire, both the new chemical analysis of urine and the new chemical doctrine of signatures were to be characterized by distillation procedures. Similarly, in a search for the ingredients of medicinal waters at spas the Paracelsians furthered the development of analytical chemistry. A long medieval tradition in this field had resulted in the development not only of isolated tests, but of real analytical procedures, and it is understandable that the Paracelsians quickly adopted this tradition and added to it. By 1571 Leonard Thurneisser (c. 1530–1596) was using quantitative methods, solubility tests, crystallographic evidence, and flame tests, and early in the next century Edward Jorden was advocating the red-blue color change of "scarlet cloth" as a regular test for those liquids that we would classify as acids and bases. The work of these men provided the basic information necessary for Robert Boyle's analytic research later in the century.

The results of the new chemical analyses were put to practical use. Chemists could now give directions for the preparation of artificial mineral waters to those who could not travel to the spas, and at the same time this analytical information added an argument for the use of chemically prepared medicines. The Paracelsians argued passionately that theirs was a new and violent age — one that had spawned ravaging diseases unknown to the ancients. (They were particularly appalled by the venereal diseases.) As a result they needed new medicines, more potent than the traditional Galenicals prepared from herbs. Their meaning was clear: These new medicines were their chemically prepared metals and minerals. The Paracelsians were not innovators in this. Nevertheless, as R. Bostocke stated in 1585, the true Paracelsian could be distinguished from others through his careful attention to dosage and his use of the chemical art to extract only the valuable essence of dangerous minerals. Furthermore, in his defense of these

Figure 2.5. Man besieged in his castle of health. From Robert Fludd, *Integrum Morborum Mysterium* (1631). From the collection of the author.

medicines (1603), Duchesne relied upon the spa-water analyses to demonstrate that minerals had beneficial medicinal effects.

The defenders of the traditional *materia medica* were far from satisfied by apologies for the chemical medicines, and, in truth, their fear of the new drugs was not groundless. Paracelsus had broken with the Galenic dictum that "contraries cure" and turned instead to Germanic folk medicine, which insisted that "like cures like." The physician was told to investigate poisons rather than bland vegetable concoctions. That poison that causes a disease should now — in proper form — become its cure. And although the chemists sought to remove the toxic qualities, the medical establishment were not reassured by this claim. For them many of the proponents of the new drugs were uneducated charlatans. In a Galenic text the very name "Paracelsian" had an unsavory connotation. Thomas Erastus (1524–1583) accused Paracelsus of advocating the internal use of lethal poisons (1572). John Donne (1573–1631), in his comparison of the innovations of Copernicus and Paracelsus, admitted only the latter to the inner sanctum of Satan's lair as the governor of the "Legion of homicide Physicians." In answer chemists spoke ever more forcefully in defense of their medicines and methods. In the mid-seventeenth century it was suggested that several hundred sick poor people be taken from the hospitals and the military camps. They were to be divided into two groups, one to be treated by the Galenists, the other by the chemists. The number of funerals would determine whether the chemical or the traditional medicine had triumphed. The trial was never made, but the fact that it was proposed indicates the heat of the controversy.

The new drugs became a subject of intense debate at the university level in the early seventeenth century. The most inflammatory pamphlets appeared at Paris in the first decade of the century, but these were soon translated and published in other parts of Europe, with histories of the conflict being written as early as 1606. In London the fellows of the Royal College of Physicians had for several decades been engaged in plans for the publication of an official pharmacopoeia. When the French chemical physician Theodore Turquet de Mayerne (1573–1655) moved to London as the physician to King James I, he increased their interest in the new chemicals. And when the pharmacopoeia did appear in print in 1618, it was seen that a careful compromise had been reached. While the bulk of the volume was devoted to the traditional Galenicals, several sections were reserved for the new chemically prepared medicines. Official sanction was given to them both there and in the preface, which called attention to their efficacy in difficult diseases.

We may then speak properly of an increasing polarization between the Hermetic physicians and the Galenists. Yet at the same time the position of the London College of Physicians shows the eventual tendency toward compromise on the difficult question of the internal use of the new medicines. Among the chemical physicians themselves an ever-increasing number sought to maintain chemistry as the basis of a new philosophy of nature, but rid it of its most mystical and least experimental aspects. Influential iatrochemists such as Daniel Sennert (1572–1637) and Andreas Libavius (1540–1616) agreed with Paracelsus that chemistry was a proper basis of medicine and was thus the chief of all sciences. But they did not wish to see the works of Aristotle, Galen, and Hippocrates discarded and burned in the marketplace. Rather than resorting to polemics, the true physician should examine both the old and the new medicines and accept the best of both. For many seventeenth-century iatrochemists the chemical philosophy could be safely followed because this seemed to provide a new observational basis for the sciences. But many of the same men were disturbed no less than the Galenists — or later, the mechanical philosophers — by the mystical, alchemical cosmology of some of their fellows. Thus, the reader of this literature will find a bewildering spectrum of medical and chemical views. These books and pamphlets encompass everything from traditional allegorical alchemy to practical chemical pharmacopoeias. And, as we shall see, the debate itself was of great concern to both physicians and scientists until well into the seventeenth century.

We might pause to reflect on the significance of the chemistry and the medical debates it engendered in this period of the late Renaissance. What had the Paracelsians accomplished? How had they influenced medicine and science in this period?

Above all, Paracelsian medicine represented a reaction against the traditional veneration for antiquity. The early Paracelsians spoke harshly of Aristotle and Galen (if not always Hippocrates) and they turned instead to the recently translated Hermetic, alchemical, and neo-Platonic texts. A vitalistic universe founded on the macrocosm–microcosm analogy and the divine office of the physician was the basis for a new Christian understanding of nature as a whole. In their drive for reform the Paracelsians proceeded to strike at the very foundations of the older system. Both the Aristotelian elements — upon which the old cosmology was founded — and their attendant humors — upon which Galenic medicine depended — were questioned. Chemists now turned to the three principles as an explanatory device, and Paracelsian physicians spoke in terms of local seats of disease governed by internal *archei* rather than the imbalance of fluids.

The Paracelsian answer to antiquity was best expressed in the emphasis on observation and experience as a new basis for the study of nature. Surely the Paracelsians were not alone in this plea, but their special interest in chemistry as a guide for the study of man and the universe distinguishes them from other Renaissance philosophers of nature. Their extensive use of chemical equipment in distillation experiments and their constant reference to chemical analogies as a means of understanding all natural phenomena place them squarely in the Hermetic–alchemical tradition.

The medicine of the Paracelsians was strongly tinged with chemistry, but not with mathematics. While they might still pay lip service to the certainty of mathematical proof, in fact their concept of quantification was closest either to neo-Pythagorean mysticism or to practical measurements by weight. Mathematical abstractions of natural phenomena and geometrical proofs savored of scholasticism, which was plainly to be avoided. Logic itself was suspect as a form of the "mathematical" science and medicine of antiquity. The medico-science of the Paracelsians thus tended to be a less rather than a more mathematicized approach to nature than that of the past.

The opinions of these chemical physicians were set forth with conviction, but often with little tact. They decried the current overreliance on antiquity. They called for a new medicine and a new natural philosophy based on chemically oriented observations and experiments. And they demanded educational reforms so that their "Christian" concept of nature might be taught at the universities. On these points they came into direct conflict with tradition. Yet they argued no less vehemently amongst themselves. Here they debated questions such as the place of mathematics in the formation of the new philosophy, the truth of the elements, the reality of the macrocosm–microcosm analogy, and the meaning of astral emanations. We can, of course, credit the Paracelsians with specific advances – their concept of disease or their recognition of the importance of chemistry for medicine (both as a basis for the understanding of physiological processes and as a new source for medicinal preparations) serve as excellent examples. And there is little question that some of the "modern" concepts of the late seventeenth century have their roots in the "nonmodern" concepts of the iatrochemists of the preceding century. Nevertheless, it was primarily by defining their vision of a new science based on medicine and interpreted through chemistry that they found themselves engaged in a debate that was to be influential in the definition of significant aspects of modern science.

CHAPTER III

The Study of Nature in a Changing World

From our brief discussion of the Paracelsians it may already be evident that it is misleading to separate the study of inorganic from that of organic nature in the Renaissance. For Aristotelians, Platonists, and Paracelsians alike in the sixteenth century the world was conceived to be alive – and at all levels. It is not unusual to read theoretical accounts of the impregnation of the earth by astral seeds and of the resultant growth of metals in veins. This process was considered by many to be comparable to the growth of a human fetus. Again, it was argued that as grains might be harvested in the fields, so too the growing metals might be harvested again and again within the earth. Such beliefs were common to Central European miners until early in the twentieth century. For Renaissance scholars there seemed little doubt that there existed in the air a spirit of life necessary to all living things. Robert Fludd's *Philosophicall Key* (c. 1619) was to explain spontaneous generation on the basis of this *spiritus mundi;* the search for the isolation of this substance was to become a major part of his life's work. Although many might have objected to Fludd's mystical bent, others would have accepted his philosophical assumptions on this point.

But, given this caveat, we still find it useful to separate the work of the miner and the metallurgist from that of the botanist, the zoologist, and the biologist. If we then turn to these fields we find that dramatic changes occurred in the sixteenth and the seventeenth centuries. Here we see medieval plant and animal lore giving way to humanistic textual criticism – and then on to the widespread search for new information through observations to replace both ancient tradition and literary criticism.

34

The Animal Kingdom

The medieval knowledge of animals derived in large part from the *Natural History* of Pliny the Elder, written in the first century A.D. Here, amidst much other information, was presented a wealth of fact and folklore relating to European, African, and Asian animals. Important to Pliny were the habits of the animals no matter how fabulous they might seem. He also described their appearance, the medical usage of their parts, and especially when they were first seen in Rome. Pliny's accounts of monsters of all sorts were repeated in the medieval bestiaries that were also bequeathed to the Renaissance scholar. But Pliny's text was no less vulnerable to humanistic criticism than that of other ancient authorities. Ermolao Barbaro (1454–1493) took on the full thirty-seven books of the *Natural History* and produced a work rivaling it in length. In his *Castigationes plinanae* (1492–1493) Barbaro rooted out errors article by article. But in typical humanist fashion he hardly concerned himself with new observations on the plants and animals described by the Roman admiral; instead he sought the ancient sources on which Pliny depended. Thus he rejected Pliny's statement that elephants lived from two to three hundred years. The correct figure – and he cited Aristotle as the source – was not three hundred but one hundred and twenty.

The encyclopedic tradition of Pliny remained fertile in the sixteenth and seventeenth centuries. The writings of Conrad Gesner (1516–1565) encompassed all aspects of knowledge, and indeed his *Bibliotheca universalis* (1545) is the first great annotated bibliography of printed books. No less significant is Gesner's *Historiae animalium* (5 vols., 1551–1621), a work that included all animals referred to by ancient and modern authorities. Included is information on each beast regarding habitat, physiology, diseases, habits, utility, and diet. Gesner included many new observations and he divided the animal world into birds, fishes, insects, and other basic categories much in the fashion of Aristotle. Within these divisions he followed an alphabetical order. Even more ambitious was Ulisse Aldrovandi (1522–1605), who published three folio volumes on birds and insects shortly prior to his death. But from his notes his students were to publish another eleven volumes – and additional manuscripts remain that have never been edited to this day.

The content of the work of Gesner and Aldrovandi was all-inclusive. While on occasion accounts of monsters were questioned, for the most part every scrap of information that could be found was presented to the reader. A reflection of this may be found in the derivative works of Edward Topsell

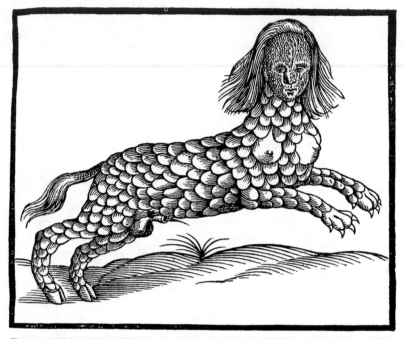

Figure 3.1. Edward Topsell's "The true picture of the Lamia." From *The Historie of Four-Footed Beastes* (London, 1607). Courtesy of the Newberry Library, Chicago.

(1572–1625), whose *Historie of Four-Footed Beastes* (1607) and *Historie of Serpents* (1608) presented to the Jacobean reader descriptions of the animals of the world in two large folio volumes. For Topsell an interest in animals was essential for the minister, who needed to correctly identify the beasts of the Bible. The same information was no less important for the physician because animals serve as food for man, because of illness resultant from their poison, and because of the medicinal use we make of their parts.

Topsell considered a simple alphabetical arrangement sufficient for his purposes. He did, however, make subdivisions under the main headings. Thus, the monstrous mantichora with its human head and triple row of teeth appears in the section on hyenas. The mantichora is only one of many mythical animals included. The unicorn is accepted because of Scriptural authority, and many other animals were included because of references to them in ancient texts (that is, satyrs, the sphinx, and dragons). One of the most unusual of these fabulous beasts is the lamia, which Topsell accepted because of a reference in the Bible (as Lilith) (Figure 3.1). Possessed of a beautiful woman's face and "very large and comely shapes on their breastes," these beasts posed a serious threat to travelers for "when they see a man, they lay open their breastes, and by the beauty thereof, entice them to come neare to conference, and so having them within their compasse, they devoure and kill them."

Topsell was also aware of animals either unknown to the ancients or those that had been rediscovered in the past century. Thus he included a number of animals both from the Americas (such as the "Land Crocodile of Brazil," actually an iguana or other lizard) and the East. Chief among the Indian animals is the rhinoceros, "the second wonder in nature," known not only from ancient authority, but also from a specimen exhibited in Lisbon (1513–1515). If he accepted the lamia, the mantichora, and dragons with little hesitation, Topsell considered the rhinoceros such a strange animal that it was necessary to assure his readers that he would not lie to them: "I would be unwilling to write any thing untrue, or uncertaine out of mine owne invention; and truth is so deare unto mee, that I will not lie to bring any man in love and admiration with God and his works, for God needeth not the lies of men."

Of special interest are the monographic studies that began to appear in the mid-sixteenth century. Chief among these were the studies of fish, birds, and marine animals written by Pierre Belon (1517–1564) and Guillaume Rondelet (1507–1566). The former toured the Near East gathering information, which he presented in his *La nature & diuersité des poissons* (1551) and his *Portraits* (1557) of animals, serpents, herbs, trees, men, and women. Belon included among "poissons" all animals living in or near the water. His inclusion of cetaceans led him to depict the birth of a killer whale with the young still attached to the placenta, thus making it possible to establish the group among the mammals (Figure 3.2). Of no less importance was Belon's essay into comparative anatomy in which he pictured side by side the skeletons of a man and a bird, drawing attention to valid homologies (Figure 3.3). He also sketched the beak of an "aquatic" bird from the new world (actually a toucan), but at the same time he was willing to give illustrations of flying serpents from the Sinai and a monstrous monk-shaped fish copied later by both Gesner and Rondelet.

The work of Rondelet was at least partially inspired by his desire to confirm the observations of Aristotle. His work comprised a careful description of Mediterranean sea life, but like Belon he included other animals associated with the water, such as turtles and seals. Nor was he averse to picturing monsters like the monk fish or the bishop fish, which he borrowed from Gesner and Belon.

Figure 3.2. The live birth of a killer whale. From Pierre Belon, *La nature & diuersité des poissons* (Paris, 1555). Courtesy of the Newberry Library, Chicago.

The late sixteenth century saw the publication or completion of a number of monographs. Gesner had asked for a book on dogs from John Caius (1510–1573) and another on insects from Edward Wotton (1492–1555) and Thomas Penny (1530–1588). The first appeared in London in 1570; the latter was put together from the notes of Wotton, Penny, and others by the Elizabethan Paracelsian physician, Thomas Moffett, and finally published in 1634.

Also significant were the increasingly detailed accounts of the flora and fauna resulting from the sixteenth-century explorations. By the end of that century pictures of some of the more striking animals were becoming fairly common in European publications, but the next century was to result in carefully prepared catalogues of animals present in the newly discovered parts of the world. Willem Piso's (1611–1678) descriptions of South American fish, birds, reptiles, and mammals gave accurate accounts of animals as exotic as the capybara, the tapir, monkeys, various sloths, the jaguar, and the South American anteaters. Jacob Bondt (1592–1631) performed a similar service for the East Indies. He corrected earlier accounts of the armorlike skin of the rhinoceros and stated that although few other Europeans had seen this beast he had seen thousands of them (Figures 3.4 and 3.5). Similarly he gave a personal account of the tiger and pictured the orangutan, which he was delighted to identify with the satyrs described by Pliny (Figure 3.6).

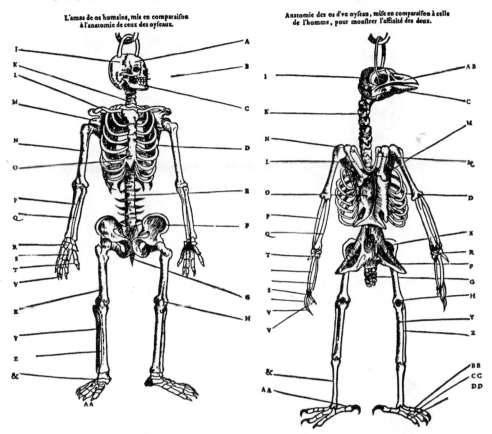

L'amas de os humains, mis en comparaifon
à l'anatomie de ceux des oyfeaux.

Anatomie des os d'vn oyfeau, mife en comparaifon à celle
de l'homme, pour monftrer l'affinité des deux.

Figure 3.3. Pierre Belon's comparison of the skeletons of a man and a bird. From *Portraits d'oyseaux, animaux, Serpens, Herbes, arbres, hommes et femmes d'Arabie & Egypte* (Paris, 1557). Courtesy of the Newberry Library, Chicago.

The Vegetable Kingdom and Medical Tradition

Medical plant lore may be traced back to a very early period, but the study of botany proper does not form a major part of the natural philosophy of antiquity. Of the botanical writings of Aristotle, only a few fragments survived; however, these did indicate the abstract nature of his interests. His pupil Theophrastus (c. 380–287 B.C.) composed a *History of Plants,* the most important sections of which refer to plant generation. This work,

Figure 3.4. Albrecht Durer's "armored" Indian rhinoceros (1515) remained the best-known illustration of the animal for nearly a century. From Edward Topsell's *Historie of Four-Footed Beastes* (London, 1607). Courtesy of the Newberry Library, Chicago.

Figure 3.5. A more realistic rhinoceros drawn from life. From Jacob Bondt's *Historiae naturalis & medicae Indiae orientalis* (Amsterdam, 1658). Courtesy of the Newberry Library, Chicago.

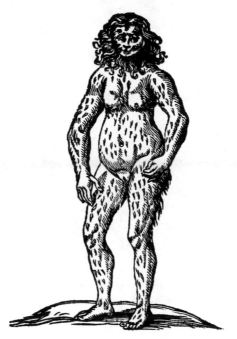

Figure 3.6. Jacob Bondt's orangutan. From the *Historiae naturalis & medicae Indiae orientalis* (Amsterdam, 1658). Courtesy of the Newberry Library, Chicago.

printed first in Latin in 1483 and then in Greek in 1497, was to become a fundamental source of the Aristotelian tradition in this field.

Yet the humanist interest in Theophrastus was surely overshadowed by the practical herbals that described plants and listed their medical uses. Here the chief survivor from antiquity was the work on *materia medica* by Pedanius Dioscorides of Anazarbeus, a military surgeon of the first century A.D. He described and illustrated some five hundred plants, emphasizing their use as drugs. The illustrations he used were partially original, but some go back to an earlier tradition that may be traced to Crateuas (first century B.C.). The high quality of ancient paintings of plants is seen best in the manuscript copy of the work in Dioscorides supposedly prepared for Juliana Anicia (early sixth century), the daughter of the Senator Flavius Anicius (Figure 3.7). Rediscovered in Constantinople in the mid-sixteenth century and sold to the Holy Roman Emperor, the illustrations would appear to have had only a limited influence on Renaissance plant illustration because the need for new paintings from live specimens had been appreciated before this time.

Figure 3.7. An example of plant illustration in late antiquity. "Stratiotes" from the Codex Aniciae Julianae (Dioscorides), c. 500 A.D. From Agnes Arber, *Herbals. Their Origin and Evolution. A Chapter in the History of Botany 1470–1670* (Cambridge: Cambridge University Press, 1912).

The history of botanical illustrations has been well documented. It has been traced from sixth-century manuscripts through the artistic decline of the Middle Ages and then on to the new union of the artist and botanist in the early sixteenth century. The strong tradition of the early illustrations makes it possible to follow the influence of individual drawings over the period of a millennium. And if details were lost as countless generations of copyists prepared new manuscripts, it remains possible to identify the originals of some of the late copies separated by hundreds of years.

The persistent interest in herbs of medical value resulted in the continued production of books dealing with them. The tenth-century herbal attributed to Macer surveyed some eighty plants and the thirteenth-century encyclopedist Bartholomew Anglicus devoted much space to plant lore. These works and others were combined with ancient tradition and local custom and resulted in a number of regional herbals that were published in the late fifteenth century. Representative of this literature is the German *Herbarius* (1485), which is filled with crude, powerful woodcuts of plants with descriptions and a listing of their medical usage (Figure 3.8). But this work goes beyond the realm of plant life. Numerous animals, including elephants, wolves, and deer, are pictured and described. Similarly metals and minerals of supposed therapeutic value (including the magnet and metallic mercury) are discussed in detail.

The early and well-illustrated editions of the German *Herbarius*, *Le Grant Herbier* (c. 1458), and the *Grete Herball* (1526) may be compared with the early printed editions of Dioscorides. Here dealing with an important early text, humanists employed their customary scholarship to produce as accurate a text as they could. As might be expected, they placed little stress on the identification of the plants described through illustrations. Accordingly, the early Greek editions of Dioscorides lack illustrations and later ones are little better. An edition as late as 1549 presented the Greek and Latin in parallel columns with an appendix listing ten pages of errors to be found in earlier editions plus variant readings. Yet, valuable though this was for the classicist, such editions were of little use for the physician.

Another difficulty, to be recognized only gradually, was that many of the plants described by Dioscorides did not exist in Northern Europe. The increasing interest in identification, the search for new medical properties, and the recognition of new plants all led to the founding of chairs of botany at European medical schools (first at Padua in 1533). The same pressures were to result in the establishment of public botanical gardens at Florence, Bologna, Paris, and Montpellier by the end of the sixteenth century.

New studies of plants and the increasing recognition of their active med-

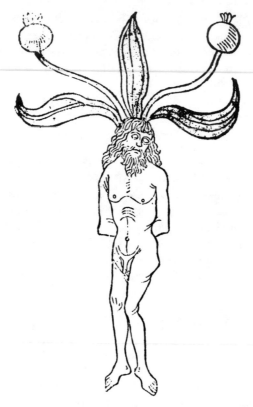

Figure 3.8. The plant mandragora from the German *Herbarius* (1485). Courtesy of the Newberry Library, Chicago.

ical properties directed ever more attention to their accurate depiction. Neither the pure texts of the humanists nor the old medieval woodcuts seemed satisfactory by the mid-sixteenth century. A new era in herbals was initiated with the books of Otto Brunfels (1489–1534) in 1530, Jerome Bock (1498–1554) in 1539, and Leonhard Fuchs (1501–1566) in 1542 (Figure 3.9). These presented new illustrations drawn from nature (Figure 3.10). Although they suffered from a number of faults (such as Brunfels's insistence on the identification of German plants with those of Dioscorides, or Fuchs's use of an alphabetical system of classification), the illustrations alone revolutionized the study of plant life. The texts were less satisfactory, and to many seemed to be no improvement on the work of Dioscorides. Thus, it was this ancient work on *materia medica* that was to remain the

Figure 3.9. Artists preparing the illustrations for Fuchs's *De historia stirpium* (Basel, 1542). Courtesy of the Newberry Library, Chicago.

most popular sixteenth-century herbal. This was largely due to the efforts of Pierre Mattioli (1501–1577), whose edition and commentary on Dioscorides (1544) updated the ancient text to include accurate illustrations and new plants discovered in the sixteenth century. This and the following century were to witness nearly a hundred editions of Mattioli's commentaries, with or without his text of Dioscorides.

. Mattioli did more than make Dioscorides a practical guide for the medical botanist. Fully aware of the new medical interest in chemical processes, he appended a description of distillation equipment and procedures to later editions of his work. For him this information seemed essential because the search for "quintessences" was by then widely recognized and because it was commonly known that such methods had not been known in antiquity. Indeed, Mattioli was here following a medieval tradition that may be seen in the work of Arnald of Villanova (c. 1235–1311), John of Rupescissa (fourteenth century), and a number of fifteenth- and early-sixteenth-century authors including Hieronymus Brunschwig (c. 1440–1512) and Philip Ulstad (fl. c. 1525).

One of the most notable of these distillation books was prepared by Conrad Gesner. His *Thesaurus Euonymi* (1555, 1569) was quickly translated into the major European languages. Much of the book was about techniques for distilling herbs and Baker, his English translator, advised his readers to

ROSA

Figure 3.10. The wild rose from Leonhard Fuchs, *De historia stirpium* (Basel, 1542). Courtesy of the Newberry Library, Chicago.

"learne the manner to separate by Arte the pure and true substance as well manifest as hidden, the which in Phisicke is a great helpe to the taking away of diseases, harde or rebellious to be cured. . . . [Then] we see plainely before our eyes, that the vertues of medicines by Chimicall distillation, are made more vailable, better, and of more efficacie than those medicines which are in use, and accustomed."

Such remedies, Baker insisted, could cure palsy, falling sickness, asthma, diseases of the spleen, the French pox, gout, dysentery, the stone, colic, and even leprosy. Baker, Gesner, and even Mattioli would have agreed that chemical distillation did make it possible to isolate effectively the pure and active portion of a medicinal herb.

Joyfull Newes Out of the Newe Found World

The inadequacy of earlier herbals was gradually made evident during the course of the sixteenth century by their outmoded illustrations, by the omission of plants common to Europe north of the Alps, and by lack of chemical information. To this list must be added the wealth of new material on plants brought to Europe by voyagers who journeyed to both the East and the West Indies. Their works told the riches of the new lands. Not only did they write of strange beasts, they also described the mineral wealth and the unusual flora. Among the greatest of these treasures were the many new herbs used as medicines by the natives. These promised new hope for supposedly incurable diseases in Europe. The descriptions of these plants appeared first in Spanish or in Portuguese but were quickly translated, abstracted, and integrated into the new herbals. Even Dioscorides was to be brought up to date with this information. Amato Lusitano (1511–1568) made it clear in his commentary (1553) that he had sought specimens of Eastern plants. And Mattioli was even more diligent in his search for new and better descriptions of Asian plants.

The chief source of Indian medicinal plants was Garcia d'Orta (1501–1568), whose *Coloquios dos simples e drogas e sonsas medicinas da India,* describing some sixty plants, was published at Goa in 1563. A Christian of Jewish ancestry, d'Orta taught medicine at Lisbon prior to sailing for Goa in 1534 to learn of the new drugs. He admitted that "if I was in Spain I would not dare to say anything against Galen and against the Greeks," but this was a new world where the authority of the ancients need no longer hold sway. Indeed, he wrote, "don't try to frighten me with Dioscorides nor Galen because I am only going to say what I know to be true." His work shows little patience with medical theory of any kind. Rather, he described diseases unknown to Western Europe (such as the Asiatic cholera) and he described the plants in use among the native physicians. These he collected and grew in his own botanical garden, which included aloes, camphor, sandalwood, and betel. In his book he identified and described the plants before going on to give an account of their pharmaceutical uses.

Hardly less important was Nicolás Bautista Monardes (1493–1588), whose first publication had attacked the use of the medicinal plants of the New World. However, his main work, *Dos libros . . . que trata de todas las cosas que traen de nuestras Indias Occidentales* (1565; two additional parts 1571 and 1574), strongly advocated the American *materia medica*. The work was soon translated into English by John Frampton (1577) under the title *Joyfull Newes Out of the Newe Found World*.

Monardes never left Spain and he was less willing to cast off the old medicine than his contemporary, d'Orta. Nevertheless, he was well aware that others had been so awed by the properties of these plants, which promised "a remedie for all maner of diseases, and hurtes," that many "have fled verie muche from the olde order and maner of Phisicke . . ."

In the pages of Monardes are to be found numerous plants and animals previously unknown to Europeans. Cocoa, sassafras, and sarsaparilla are discussed in detail, and mechoacan, the "Rhubarb of the Indies," now known to be simply a mild purgative, is offered as a cure for a wide range of ailments. Guaiac wood is offered as the true Indian cure for venereal disease and tobacco is described at great length as a medicine, along with numerous derivative ointments and mixtures (Figure 3.11). He noted, however, that the Indians "take the smoke of the *Tabaco,* for to make theim selves drunk withall and to see the visions, and things that doe represent to them, wherein thei doe delight."

The works of both d'Orta and Monardes were widely popularized throughout Europe by Charles l'Ecluse (1526–1609), who translated and epitomized them in Latin. Translations into other languages followed rapidly. John Frampton's English translation has already been noted; the French Paracelsist Jacques Gohory (1520–1576) was one of the first to emphasize the broad curative powers claimed for mechoacan by Monardes.

Added to this was the constant influx of new information. Juan Fragoso (sixteenth century) described the aromatic substances and fruit trees common to India in a book printed in 1572. Christavão da Costa's (c. 1540–1599) *Tractado de las drogas y medicinas de las Indias Orientales* (1578) was partially derived from the earlier text of d'Orta, but it contained much new material as well as illustrations he had made himself. Hardly less important was Jacob Bondt's *De medicina Indorum* (1642), describing the plants and diseases of the East Indies with a new thoroughness. Bondt's work became widely known not only on its own, but also by being published together with a number of editions of Prospero Alpini's (1553–1617) book on Egyptian medicine (1591) in the mid-seventeenth century. In his *De plantis Aegypti* (1592) Alpini went on to describe fifty-

Figure 3.11. The tobacco plant. From Nicholas Monardes, *Joyfull Newes Out of the Newe Found World*, trans. John Frampton (London, 1577). Courtesy of the Newberry Library, Chicago.

seven Egyptian plants. For North America Thomas Hariot's (1560–1621) *A Briefe and True Report of the New Found Land of Virginia* (1588) offered the reader some extraordinary engravings, but failed to present the detailed information being gathered by Spanish, Dutch, and Portuguese explorers, naturalists, and physicians.

Observation and Order

There is little doubt that herbals were among the most popular books printed in the sixteenth and seventeenth centuries. One need not spend much time with them to become convinced of the rapidly expanding knowledge in this field. Building upon Dioscorides (some five hundred plants described) and a medieval tradition (eighty plants described in Macer), Renaissance botanists quickly found thousands of plants previously either unknown or never properly described in the past. It was necessary to illustrate them, to prepare accurate descriptions, and above all to gather information regarding their medicinal properties. Mattioli and others radically altered Dioscorides to make this ancient text useful in a new age. But others were not so satisfied and a flood of new herbals were offered to the public. There are so many of these that there would be little point in attempting to make a complete list, but note should be taken of William Turner's (c. 1510–1568) herbal (1551–1568) that was filled with many new observations and careful descriptions (particularly of English plants), of Rembert Dodoens' (1517–1585) *Pemptades* (1583) with its nearly nine hundred illustrations, and of Mathias Lobelius' (1538–1616) many books of plants. In England John Gerard's (1545–1612) *Herball* of 1597 remains of interest for its detailed description of English garden plants as well as the tomato and the "Virginian" potato, but it drew heavily from Henry Lyte's (c. 1529–1607) 1578 translation of Dodoens. Gerard, in turn, served as a basis for John Parkinson's much larger *Paradisus* of 1629. In all of this activity it was general practice to borrow freely both illustrations and descriptions from any work at hand.

The most comprehensive works were composed by the Bauhin brothers. Jean Bauhin's (1541–1613) *Histoire universelle des plantes* (published posthumously, 1651) described five thousand plants and included thirty-five hundred illustrations. His brother Gaspard (1560–1624) was even more industrious, and his famed *Pinax* (1623), including information on six thousand plants and claiming six hundred entirely new descriptions, was to remain a fundamental source for botanists for the next two hundred years.

The quantity of the new information resulted in an organizational problem hardly dreamed of by medieval or early-sixteenth-century herbalists. They had been concerned primarily with the need to point out the medical properties of plants. At that time few authors were troubled with the problem of classification. For many an alphabetical arrangement sufficed and this still seemed satisfactory for Leonhard Fuchs (1542) and William Turner (1568). John Parkinson (1629) divided plants into "sweet smelling"; purging; venomous, sleepy, and hurtful and their counterpoisons; wound herbs; cooling; hot and sharp; thistles, and so on to a total of seventeen categories. Not knowing what to do with some that did not seem to fit anywhere, he added one category entitled "the unordered tribe."

If Fuchs remained satisfied with his traditional alphabetical arrangement, his contemporary, Jerome Bock (1539), was not. Following Aristotelian tradition he divided his material into herbs, shrubs, and trees, but he noted that

"I have placed together, yet kept distinct, all plants which are related and connected, or otherwise resemble one another and are compared, and have given up the former old rule or arrangement according to the A.B.C. which is seen in the old herbals. For the arrangement by the A.B.C. occasions much disparity and error."

The Bohemian botanist Adam Zaluziansky von Zaluzian (1558–1613) discarded earlier systems to arrange his *Methodi herbariae* (1592) in a novel fashion, beginning with the simplest plant life and then going on to more complex forms. His work has added interest in that he argued for the separation of botany from medicine.

"It is customary to connect Medicine with Botany, yet scientific treatment demands that we should consider each separately. For the fact that in every art, theory must be disconnected and separated from practice, and the two must be dealt with singly and individually in their proper order before they are united. And for that reason, in order that Botany (which is, as it were, a special branch of Physics) may form a unit by itself before it can be brought into connection with other sciences, it must be divided and unyoked from Medicine."

Such a statement could hardly have been made a century earlier.

The problems of classification occupied many sixteenth- and seventeenth-century botanists and zoologists, and a number of schemes were presented. Lobelius suggested leaf form as a basis for classification; this was rejected by Fabio Colonna (1567–1650), who argued that other parts of the plant — such as the flower, the receptacle, and the seed — were essential for any such scheme. Andrea Cesalpino (1519–1603), while seeking to reestablish the authority of Aristotle in the study of natural philosophy, prepared his *De plantis* (1583), in which he advocated a scheme based upon flowers and fruit.

Gaspard Bauhin, partially influenced by Cesalpino's arrangement, used a binomial form of nomenclature for plants. Drawing on common characteristics he divided the *Pinax* into twelve books and these, in turn, were broken down into sections. The former correspond very roughly to our *genera*, the latter, to *species*. The plants included within the sections depend upon common properties. Here Bauhin was sometimes successful — as when he assembled one group of plants sharing narcotic (chemical) properties — and at other times less so — as when he placed together an unusual group that had little in common with one another except that they all yielded useful spices. Like Zaluziansky he progressed from simpler forms of plant life (grasses) to more complex ones (trees).

The work of Bauhin was developed by Joachim Jung (1587–1657) and John Ray (1627–1705). The latter arranged both plants and animals into systematic groupings — the remnants of which are still evident in current classification. Thus, although the work of Carolus Linnaeus (1707–1778) remains the foundation of modern classification of both plants and animals, the problems presented by the vastly increased number of known types had resulted in over a century of efforts at systematization upon which he had been able to build his own work.

Perhaps in no sphere of scientific history are the changes of the sixteenth and seventeenth centuries more evident to the eye than in botany and zoology. As we follow the early printed books we first find the fifteenth-century herbals, which still reflect the medieval world with their crude woodcuts of plants and animals as well as with their folklore. The influence of Renaissance humanism is seen in Barbaro's *Castigationes* of Pliny and the efforts to modernize Dioscorides and in the careful attention to textual criticism. But for all the pains taken by humanist scholars on this score, neither the physician nor the botanist could benefit from such studies until illustrations and descriptions of plants were improved. This was the contribution of the German fathers of botany: Brunfels, Bock, and Fuchs and their followers in the late sixteenth century. The greater accuracy was accompanied by a vast increase in the numbers of plants known. This development is due partially to the new interest in European plants, and partially to the fascination with the flora and fauna of the newly discovered parts of the world. The original five hundred plants in Dioscorides had swollen to six thousand in Bauhin's *Pinax* of 1623. And if the organization of a few hundred plants seemed a problem of little concern to the earlier herbalists, it led to debates on classification by the turn of the new century — debates that were not settled until well into the eighteenth century.

The growing knowledge of the vegetable kingdom was paralleled by that of animals. In the course of the sixteenth and the seventeenth centuries the scattered information to be found in the medieval herbals and the tales related by Pliny and the old bestiaries gave way to the encyclopedic studies of animals by Gesner and Aldrovandi. These efforts were supplemented by monographs on birds, fishes, insects, and other animals prepared by authors who observed from nature and tried to divorce themselves from the accounts of antiquity. Here, as in the case of plants, enthusiasm about new life forms was greatly fostered by European voyages to the Americas, Asia, and the East Indies.

But if these crucial centuries witnessed a veritable knowledge explosion unequaled earlier, it would be wrong to interpret this simply in terms of a triumph of modern science. The work of Edward Topsell indicates the deep-rooted belief in mythical beasts whereas the works of Gesner, Belon, and Rondelet all include monstrous forms along with living species. Perhaps the best reflection of the period may be seen in Bondt, who found in the orangutan proof of the existence of the satyrs of antiquity. Indeed, the search for monsters of all kinds remains in evidence in the pages of the early issues of the *Transactions* of the Royal Society of London late in the seventeenth century.

In the herbal tradition we might have recognized a continuation of the old doctrine of signatures. According to that doctrine the correspondence of the name or shape of a plant to that of a human organ indicated the plant's proper medical usage. Here Paracelsians sought reform not through denying the doctrine in principle, but rather through the introduction of chemical methods. True identification of a "sign," they argued, was only possible by distillation, which would separate the pure essence of the plant from its gross outer substance. Beyond this, one may observe that the connection of astrology with plant lore was surely not declining in the period we have touched upon.

A similar growth of knowledge may be followed in the Renaissance study of the human body, the subject of the next chapter. But as we shall see, there too is to be found a connection of mysticism and rigorous observational technique.

CHAPTER IV

The Study of Man

A new understanding of the human body in the Renaissance resulted from anatomical studies pursued with an intensity probably unequaled at any time earlier. Here we may follow a series of Paduan professors and their students: Andreas Vesalius, Realdo Columbo (c. 1510–1559), Gabriele Falloppio (1523–1562), Hieronymus Fabricius of Aquapendente (c. 1533–1619), and William Harvey. But the period between the publication of the *De fabrica* (1543) of Vesalius and the *De motu cordis* (1628) of Harvey also witnessed the contribution of others whose work does not always fit so neatly into the pattern of modern science. Thus, Michael Servetus (c. 1511–1553) described the pulmonary transit of the blood in a theological tract (1553), and we shall find that the analogy and interrelation of the macrocosm and the microcosm reappears here again as a stimulus for both research and speculation. Perhaps nothing is more revealing of the complexity of early-seventeenth-century intellectual currents than the fact that most of the main figures of this story openly professed their allegiance to Aristotle and Galen whereas the English alchemist-physician Robert Fludd was the first to support the Harveyan circulation in print (1629) because of what seemed to him to be its deep mystical connotation.

The Medieval Heritage

As in all other fields of science, the anatomy and physiology of the Renaissance was based initially upon texts and concepts that had survived from earlier periods. In part this meant an acceptance of the macrocosm-microcosm analogy as understood by Aristotle and it also meant a vitalistic view of nature — one that would become the object of attack by seventeenth-century mechanists.

But in addition to the survival of ancient philosophical concepts, there was also an imposing body of anatomical and physiological information that had come down from antiquity. Both Alcmaeon of Croton (c. 500 B.C.) and Aristotle (384–322 B.C.) had been interested in the description of the parts of men and animals, and in Alexandria Herophilus and Erasistratus (fl. 280 B.C.) presided over a flourishing school of anatomy. Later it was said that they had made their observations through the vivisection of criminals furnished to them by the Ptolemies, but whether or not this is true, there is little question that the Alexandrian anatomists did dissect human cadavers. The results were to be seen in detailed anatomical descriptions of the parts of the human body and their comparison with the similar parts of animals. In addition, their physiological investigations were to be the subject of discussion for nearly two thousand years.

If the Alexandrian anatomical studies remain a high point of ancient science, their full texts were lost in succeeding centuries as a result of the overwhelming influence of Galen. Originally a surgeon to the gladiators in his native city, Pergamum, Galen traveled widely throughout the empire and was to write on all aspects of philosophy and medicine. It was his work rather than that of his predecessors that was to be abstracted and digested by the medical authors of late antiquity and Islam. And it was the continued influence of his work that Paracelsus and his chemical followers reacted against in the sixteenth century. It would be hopeless to try to summarize the extensive writings of Galen, but it is necessary to note a few points that proved particularly vexing to sixteenth-century anatomists.

Galen's anatomical and physiological works were both lengthy and detailed. His work is of particular importance for his examination of the spinal cord, the mechanism of breathing, and the cardiovascular system. But his conclusions were based only to a small degree upon human dissection. He relied most upon animals that were easily available: that is, sheep, oxen, pigs, dogs, and especially the Barbary ape. It is not surprising, then, that he should have made some notable errors. A five-lobed liver (based upon the dissection of a dog) and the *rete mirabile* (an elaborate network of blood vessels not present in man) were described as part of human anatomy. These errors and others were to remain part of anatomical teaching until the sixteenth century.

Of special significance was Galen's description of the cardiovascular system (Figure 4.1). Here the discovery of fundamental errors in his texts during the Renaissance was to lead to a radically new concept of the blood flow. For Galen the blood originated in the liver and from there proceeded to all parts of the body through the veins. Rich in natural spirits, this ve-

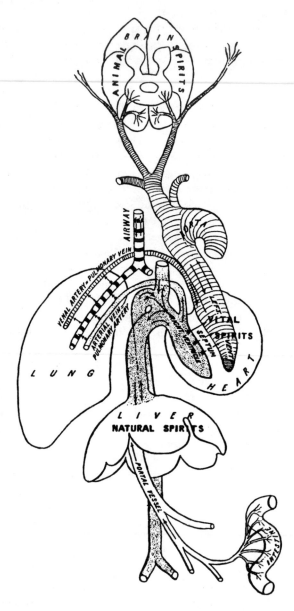

Figure 4.1. Diagram of Galen's physiological system. From Charles Singer, *The Discovery of the Circulation of the Blood* (London: Wm. Dawson and Sons, Ltd., 1956). Courtesy Mr. A. W. Singer.

nous blood had the function of nourishing the body tissues and at the same time carrying away waste substances. Eventually this spent venous blood reached the right ventricle of the heart. Most of it was routed first to the lungs and then to the liver after having been purged of accumulated impurities. However, Galen postulated the existence of pores between the right and left ventricles of the heart through which a very small portion of the venous blood moved to the left cavity. This portion was there combined with air from the lungs to form the vital spirits required for life and these were then distributed through the arteries. A final transformation occurred in the brain, where animal spirits were prepared and routed through the nerves. The key to this system rested in the interventricular pores, openings that do not exist. When this was discovered a rethinking of the entire system was necessary.

But the revision of Galenic physiology was not to take place for more than a millennium. This was due partially to the fact that Galen was the last great figure of Greek anatomy and physiology. But it was due also to the fact that later ancient physicians applied themselves less to new research than to the abridgment and codification of Galen's writings. And if Islamic medicine was later to be strongly influenced by Galen, the emphasis of the Arabic texts was directed more toward the cause and cure of disease than to anatomy and physiology. The learned world of the West was first to reflect the Eastern texts from translations made in the thirteenth century. And because of these Islamic interests, Western scholars of the Middle Ages knew relatively little of the anatomical works of Galen. Only an abridged version of *On the Use of the Parts* was readily available to thirteenth-century physicians.

In contrast to the limited number of anatomical texts available, a good omen for the future was to be found in the early revival of public dissections. Animals were examined once more at Salerno in the thirteenth century, and early in the fourteenth century Bologna became the center for anatomical studies. Here the stimulus originated not from the medical faculty, but from the law school, whose members saw the need for postmortem examinations. The anatomical text prepared there by Mondino de' Luzzi (c. 1275–1326) in 1316 was to become the standard for public dissections until well into the sixteenth century (Figure 4.2). In it Mondino proceeded to describe first the organs of the abdominal cavity and then worked outward to the head and the extremities. This order gave first attention to those parts most likely to decay, a subject of great importance in a period that lacked adequate preservatives.

Medieval medical schools were quick to recognize the significance of

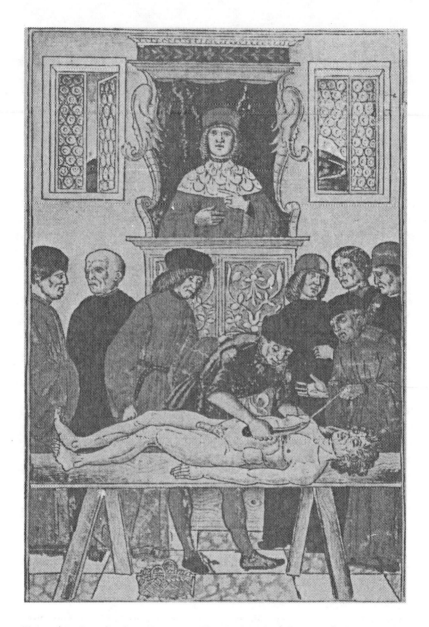

Figure 4.2. A medieval anatomy scene. The physician-lecturer expounds the text – most likely Mondino – while the barber-surgeon dissects the cadaver. From the *Fasciculo de Medicina* (1493), as reproduced in *Medical World News* (12 January 1976).

such demonstrations, and it soon became mandatory for medical students to attend a specified number of public anatomies. By 1400 the public anatomy had become a standard part of the curriculum at most universities. Yet it must be stressed that these were demonstrations meant to acquaint students with the parts of the body; they made no pretense of being new research.

An Anatomical Renaissance

The persistence of tradition ensured the publication of Mondino's anatomy in the late fifteenth century, but the medical humanism of the period placed a new stress on the texts of antiquity. It was soon recognized that the most important anatomical works of Galen had not been available to scholars, and a determined effort was soon underway to prepare them for publication both in Greek and Latin. Galen's *On the Use of the Parts* appeared in several Latin translations by 1500, and the English medical humanist, Thomas Linacre, founder of the London College of Physicians (1518), dreamed of the publication of a complete Greek edition of Galen. He was in fact responsible for a number of editions of individual medical works, among them Galen's *On the Natural Faculties* (1523). Even more industrious was Johannes Guinter of Andernach, professor of medicine at Paris, who devoted much of his early professional life to the preparation of Greek medical texts. Not only did he translate the larger part of Galen, but he also prepared editions of a number of other physicians of late antiquity: Paul of Aegina (late seventh century), Caelius Aurelianus (seventh century), Oribasius (c. 325–c. 400), and Alexander of Tralles (sixth century). In addition, he wrote books on the plague, medical spas, obstetrics – and, in his old age, a defense of Paracelsian chemical medicines.

Of real significance is the fact that Guinter had only recently completed his translation of Galen's *On Anatomical Procedures* (1531) when a new student, Andreas Vesalius, entered the medical program at Paris. Recognizing the talent of this young man, Guinter enlisted him as an assistant in the preparation of his own text, the *Anatomical Institutions According to the Opinion of Galen for Medical Students* (1536). Although Vesalius was later to express a low opinion of his master's expertise, the fact remains that he gained from him the most advanced training in this field available anywhere in Europe – plus Guinter's strong Galenic bias.

A shrewd observer, Vesalius quickly became aware of many errors in Galen – and the need to depict the parts properly. Here his work may be seen as analogous to that of Brunfels, Bock, and Fuchs, who realized the importance of new illustrations in the field of botany. But it would be wrong to credit Vesalius alone with modern anatomical illustration. Leonardo da Vinci's (1452–1519) earlier anatomical drawings were masterful, but unfortunately of little impact as they were unpublished. But others did anticipate Vesalius on this score, most notably Berengario da Carpi of Bologna (c. 1460–c. 1530), who had new anatomical drawings prepared for his commentary on Mondino.

The path to the *De humani corporis fabrica* was a relatively short one. Leaving Paris, Vesalius taught for a year at Louvain (1536) and then took his medical degree at Padua, where he was immediately appointed lecturer in surgery (1537). In addition to his travels and teaching Vesalius was constantly writing. In 1538 his *Tabulae sex* — six sheets of text and illustrations — appeared at the request of his anatomical students. Three of these were illustrated by Jan Stephen van Calcar (1499–c. 1550), a student of Titian. In 1541 Vesalius contributed to an edition of Galen, and two years later his masterwork on the human body was published (Figure 4.3). Shortly after its publication he was appointed physician to the emperor, Charles V, and then to his son, Philip II of Spain. A man of great energy, Vesalius was preparing to return to Padua when he died while returning from a pilgrimage to Jerusalem in 1564.

The *De fabrica* had its greatest impact because of its plates (possibly also by Jan Stephen van Calcar) (Figure 4.4). But when we turn to the text we find the expected Galenic foundation. Like other medical humanists Vesalius avidly sought minor errors in the ancient texts. This was accepted scholarship and did not affect the general esteem in which the ancient physicians were held. But Vesalius adhered to a Galenic organization as he led the reader first to the skeleton, then the muscles, the cardiovascular system, and finally the brain and the organs of the abdominal and thoracic cavities. This order is the reverse of Mondino's practical method of dissection.

Our own interest must remain centered on Vesalius's treatment of the heart and the arterial and venous systems. Here he was decidedly conservative. The Galenic spirits (natural, vital, and animal) are not questioned and there is no real departure from Galen in his discussion of the blood flow. As to the heart, Vesalius was so reluctant to reject the Galenic position regarding the interventricular pores of the septum that he wrote that

"The septum is formed from the very densest substance of the heart. It abounds on both sides with pits. Of these none, so far as the senses can perceive, penetrate from the right to the left ventricle. We wonder at the art of

Figure 4.3. The title page of the *De fabrica* (Basel, 1543). Note that it is Vesalius himself who is dissecting the body — a far different scene from the Mondino illustration of 1493. Courtesy of the Newberry Library, Chicago.

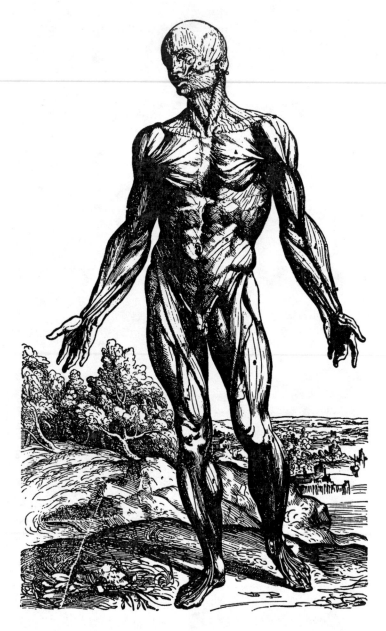

Figure 4.4. One of the plates illustrating the human muscles. From Andreas Vesalius, *De humani corporis fabrica* (Basel, 1543). Courtesy of the Newberry Library, Chicago.

the Creator which causes blood to pass from right to left ventricle through invisible pores."

Here observation is rejected in favor of authority, which affords us some insight into the authority Galen commanded in the mid-sixteenth century.

Vesalius was to publish a second edition of the *De fabrica* in 1555. Here he returned to the problem of the interventricular pores of the septum once more. But now he observed that "although sometimes these pits are conspicuous, yet none, so far as the senses can perceive, passes from right to left ventricle." Galen is now rejected, but the reader senses that it was a painful decision.

"Not long ago I would not have dared to turn aside even a hair's breadth from Galen. But it seems to me that the septum of the heart is as thick, dense and compact as the rest of the heart. I do not see, therefore, how even the smallest particle can be transferred from the right to the left ventricle through the septum."

This decision, confirmed by contemporary anatomists, was to result in a complete rethinking of the blood flow.

Vesalius himself was concerned about the criticism he received from those more conservative than he was himself: "I surely deserved something better than the slander of those who are so furiously aroused against me because their studies have not been so successful, because I don't accept Galen, and because I refuse to disbelieve my eyes and reason for his sake." Much as he revered the works of Galen himself, he rejected pointed suggestions that he publish only a work of Galenic annotations in the more traditional humanist fashion.

The Blood Flow: Vesalius to Harvey

Vesalius may not have been as much an innovator as once suggested, but there is no doubt that the *De fabrica* almost immediately became the most influential of all anatomical texts. The book and its illustrations were copied, and evidence exists to indicate its importance in a succession of master-pupil relationships that begin with Vesalius and his replacement at Padua. His successor was Realdo Columbo, who had been his assistant. He, in turn, was succeeded by Gabriele Falloppio, who is famed for his study of the human oviduct and after whom are named the Fallopian tubes. Fallopio was followed by Hieronymus Fabricius, whose work on the valves in the veins was well known to his student William Harvey. This remarkable succession indicates the close and direct connection between Vesalius and Harvey.

The work of Vesalius had established the structure of the heart, but it surely had not settled its function. The basic physiology remained Galenic and derived from the *On the Use of the Parts*. Indeed, in regard to the heart, the lungs, and the arteries, there were few changes from Galen.

However, the followers of Vesalius almost immediately began to go beyond him in their study of the interrelation of these parts and in their study of the purpose of respiration. The first to do so was Michael Servetus, whose early training bears some similarity to that of Vesalius. True, he was to become something of an astronomer, a mathematician, and a theologian, but he was also trained as an anatomist at Paris under Guinter. Like Vesalius, Servetus was to act as Guinter's assistant. Once again we find a Galenist, which may be seen in his *Account of Syrups . . . According to the Judgment of Galen* (1536). But another early medical work attacked Leonhard Fuchs as a heretic, and there can be little doubt that Servetus was himself a religious radical in an age when it was dangerous to be a nonconformist. A Unitarian who could stomach neither Catholic nor Protestant, Servetus had published two books on the errors of the doctrine of the Trinity before the age of twenty-one. A hunted man, he assumed the name of Villanovus and took employment with a publisher. By 1546 he had written his *Restitution of Christianity,* which he sent in manuscript — along with another work on the Trinity — to John Calvin in Geneva. Now forced to flee Catholic France, he began a journey to Italy by way of Geneva. There he was recognized as the man who had attacked Calvin and was arrested while attending church. Tried for his errors, he was sentenced to death and burned at the stake with his book in 1553.

Of the thousand copies of the *Restitution of Christianity* printed in 1553 only three have survived. The book itself is an expression of Servetus's religious convictions, but it is of importance to us primarily for its fifth chapter on the Holy Spirit and the dispensation of the Divine Spirit to man. Here Servetus discussed the respiration and the relationship of spirit and air. For him the vital spirit in the body resulted from a mixture of subtle blood and inspired air — and this was not formed in the left ventricle of the heart as postulated by Galen, but rather in the lungs. Rejecting the Galenic concept of a "sweating" of blood from the right to the left ventricles, Servetus correctly described the pulmonary transit: that is, that blood from the right ventricle is pumped through the pulmonary artery to the lungs. Here a color change occurs in which the venous blood becomes lighter due to the inspired air. From there the blood travels to the left ventricle via the pulmonary vein and is then distributed through the arterial system. It is now known that this portion of the blood flow had been correctly described by Ibn an Nafis in the thirteenth century, but there is no indication that any sixteenth-century anatomists knew his work.

It is important to insist on taking the description of Servetus in context. Here as elsewhere in this period it may be noted that significant observations occur in contexts that would be considered nonscientific today. Similar considerations regarding the aerial spirit and its relationship to the blood were to lead Robert Fludd to describe a mystical circulation of the blood just seventy years later — and were also to lead him to compose the first defense of Harvey's *De motu cordis.*

Although Servetus's views on the pulmonary transit were known a century later, it is doubtful whether the work was influential in the mid-sixteenth century. Nevertheless, there was a rapid acceptance of a nonporous septum in the last half of the century, and this fact required a new explanation of the origin of the arterial blood. Realdo Columbo reached the same result as Servetus — and probably independently — in his text on anatomy (1559). Here — in contrast to Servetus — we feel that we are in the solid tradition of Vesalian anatomy. Again we see the Galenic influence, for it is the liver that perfects the nutrient blood that is then distributed throughout the body. Similarly it is the left ventricle that distributes the vitalized blood through the aorta. However, Columbo's precise description of the heart and its valves requires the pulmonary transit for the passage from right to left. After Columbo this "lesser circulation" was generally accepted.

The final connecting link between Harvey and the school of Padua may be found in the work of Harvey's teacher in anatomy, Hieronymus Fabricius, who was associated with that university for sixty-four years. While Fabricius may be most frequently cited for his work on comparative anatomy, he is particularly important as a predecessor of Harvey for his description of the valves of the veins. His research on this subject was carried out while Harvey was a student at Padua and was published in 1603. But Fabricius, a thorough Galenist, insisted that these valves existed in order to prevent dangerous dilations. The valves acted to prevent the starvation of other parts should one locality require excess nutriment. Their absence in the arteries was explained by the normal ebb and flow of the arterial blood. For Harvey the valves were to serve as the proof of a one-way circulation of the blood.

Combined with accelerating anatomical knowledge was a new tendency at the end of the sixteenth century for some scholars to speak in general terms of a circulation of all the blood in the body. There were a number of reasons for this. Some — like Cesalpino — were militant Aristotelians. Aristotle had spoken of the primacy of the heart in the body, and Cesalpino was to adopt the pulmonary transit as it seemed to confer greater importance to the heart. Others thought in terms of mystical celestial influences conforming to correspondences between the macrocosm and the microcosm. This accepted, it seemed only right that a microcosmic circuit should imitate the planetary (or solar) revolutions. Indeed, Roch le Baillif wrote a work on man and his "essential anatomy" that emphasized analogies between the two worlds rather than the physical anatomies of the schools. And finally there were Paracelsians who suggested that the parts of the body acted as pieces of chemical equipment. In this case it was argued that the blood circulates in the body in a manner similar to distillation "circularity." If this was all speculation, it does show the trend of the times and indicates something of the breadth of thought regarding the blood flow in the period immediately prior to Harvey.

William Harvey and the Circulation of the Blood

Like so many major figures in Renaissance science, Harvey built upon the work of his predecessors and brought together a number of seemingly disparate themes. Educated first at Cambridge, Harvey traveled to Padua in 1597 where he studied under Fabricius when that teacher was preparing his work on the valves of the veins. After taking his medical degree, Harvey returned to England in 1602 where he became a physician at St. Bartholomew's Hospital and Physician Extraordinary to James I. Elected a Fellow of the Royal College of Physicians (1607), he was a member of one of the most prestigious medical (and scientific) associations in Europe. He himself was to give the Lumleian Lectures there on anatomy (1615), the notes of which indicate to us his early interest in the subject of the blood flow.

Fundamental to our understanding of Harvey was his Paduan education. Because of this training, he was an admirer of both Aristotle and Galen. The extent of this may be seen in the De motu cordis (1628), where Harvey seems more than willing to ascribe the discovery of the pulmonary transit to Galen, and in his discussion of scientific method in the De generatione animalium (1651), which is based closely on the Analytics and Physics of Aristotle.

But the discovery of the circulation was based upon more than reverence for ancient genius and the belief that it was their work that should properly serve as the basis for a new scientific age. Harvey's work echoes the contemporary interest in new observations, in mystical analogies, and even in the use of mechanical examples.

The *De motu cordis* is a small volume, but one that displays a thorough knowledge of anatomical literature as well as Harvey's own observational evidence. Turning first to the heart itself, Harvey had examined it and the motion of the blood in some forty species. He observed that in all cases the heart hardens as it contracts and that as this contraction occurs, the arteries expand. The periodic expansions could be felt at the wrist as the pulse and he rightly assumed that this occurred because blood was being pumped into the arteries. The action of the heart then, Harvey observed, might be compared with that of a water bellows.

Referring to the hearts of cold-blooded animals because of their slower action, Harvey noted that the auricles contract first, and then the ventricles. The process is carefully described. First the blood enters the right auricle through the vena cava. As this contracts, the blood is transferred to the right ventricle, where valves make it impossible to flow backwards. The right ventricle contracts next, sending the blood through the pulmonary artery to the lungs. Again valves make it impossible to reverse direction. And, because there are no pores in the septum, all this blood is sent through the lungs. On the left side of the heart, blood from the lungs first enters the left auricle from the pulmonary vein. Then as this vessel contracts, the blood proceeds to the left ventricle. Further contraction forces the arterial blood into the aorta and the arterial system.

This was all of great interest as a physiological discovery, but Harvey now went further. Reflecting on the valves of the veins, he stated that the blood flow proceeded not only in one direction to the heart, but continuously in one direction throughout the body (Figure 4.5). At this point Harvey advanced a telling quantitative argument. Assuming that the left ventricle held only 2 ounces of blood, and that the pulse beats 72 times a minute, then in one hour the left ventricle would force into the aorta some 540 pounds of blood. At most, animals have only a few pounds of blood in their bodies and one must ask where this blood comes from — and where does it go? Harvey's conclusion is that the blood sent out from the aorta could only come from the veins:

"I began to think whether there might not be a motion, as it were, in a circle. Now this I afterwards found to be true; and I finally saw that the blood, forced by the action of the left ventricle into the arteries, was distributed to the body at large, and its several parts, in the same manner as it is sent through the lungs, impelled by the right ventricle into the pulmo-

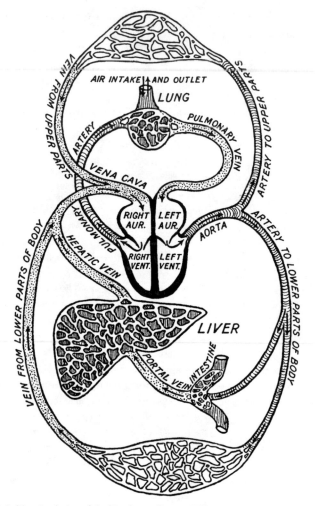

Figure 4.5. The circulation of the blood according to William Harvey. From Charles Singer, *The Discovery of the Circulation of the Blood* (London: Wm. Dawson and Sons, Ltd., 1956). Courtesy Mr. A. W. Singer.

nary artery, and that it then passed through the veins and along the vena cava, and so round to the left ventricle in the manner already indicated. Which motion we may be allowed to call circular, in the same way as Aristotle says that the air and the rain emulate the circular motion of the superior bodies; for the moist earth, warmed by the Sun, evaporates; the vapors drawn upwards are condensed, and descending in the form of rain, moisten the earth again . . ."

The significance of the heart, evident throughout, takes on a still greater role when undertood in cosmic terms through the macrocosm–microcosm analogy:

"The heart, consequently, is the beginning of life; the sun of the microcosm, even as the sun in his turn might well be designated the heart of the world; for it is the heart by whose virtue and pulse the blood is moved, perfected, made apt to nourish, and is preserved from corruption and coagulation; it is the household divinity which, discharging its function, nourishes, cherishes, quickens the whole body, and is indeed the foundation of life, the source of all action."

Turning finally to the observations of the valves in the veins, Harvey showed that the flow of blood is always toward the heart and never away from it. His conclusion was thus well founded, for

"Since all things, both argument and ocular demonstration, show that the blood passes through the lungs and heart by the action of the [auricles and] ventricles, and is sent for distribution to all parts of the body, where it makes its way into the veins and pores of the flesh, and then flows by the veins from the circumference on every side to the centre, from the lesser to the greater veins, and is by them finally discharged into the vena cava and right auricle of the heart, and this in such a quantity or in such a flux and reflux thither by the arteries, hither by the veins, as cannot possibly be supplied by the ingesta, and is much greater than can be required for mere purposes of nutrition; it is absolutely necessary to conclude that the blood in the animal body is impelled in a circle, and is in a state of ceaseless motion; that this is the act or function which the heart performs by means of its pulse; and that it is the sole and only end of the motion and contraction of the heart."

The Verdict

Although the discovery of the circulation of the blood is today judged to be the only physiological achievement of the early seventeenth century comparable to the contemporary developments in the physical sciences, the immediate reaction to the *De motu cordis* was a mixed one. But if there were a number of conservative authors who rejected the book flatly, it is of more interest to note the spectrum of views of those who supported Harvey.

The first to defend Harvey was his friend and colleague, Robert Fludd. Long interested in an aerial life spirit and its assimilation in man's body, Fludd had described a mystical circulation of the arterial blood as a necessary consequence of the macrocosm–microcosm analogy in 1623. A witness to Harvey's anatomical demonstrations, he had arranged to have his friend's book published by his own publisher in Frankfurt. In composing his own book on the pulse (1629), Fludd spoke of his "esteemed compatriot" who had confirmed the circulation both with philosophical arguments and observational demonstrations.

But for Fludd it is evident that Harvey's anatomical evidence simply confirmed deeper mystical truths. This was understood correctly by Marin Mersenne (1588–1648) and Pierre Gassendi (1592–1655), contemporaries of Descartes and Galileo. In a detailed reply to Fludd (1631) Gassendi attacked both Fludd and Harvey on the circulation, the reason being that Gassendi professed once to have seen the presence of the pores in the septum demonstrated. And if they were there, he argued, they had to have a purpose. The proper purpose was the formation of the arterial blood, and therefore the Galenic system of the blood flow could be maintained. Not so, replied Fludd (1633); Harvey's dissections had proved to him the impermeability of the septum, and the values of the heart indicated the blood flow leading from the right to the left side of the heart through the lungs. For him the circulation was a fact, but one that could, and had been, postulated by him on the basis of cosmic truths prior to Harvey's lesser – but no less convincing – physiological evidence.

If Harvey's work could be interpreted as a proof of mystical cosmology by Fludd and attacked by mechanists for this same reason, we need not be surprised to find that René Descartes also had some reservations about Harvey's explanation. In his *Traité de l'homme* (1632) he accepted the total circulation of the blood, but only by explaining the heart in terms of a mechanical distillation unit. Assuming a higher than bodily temperature of the heart and a cooling action inherent in the lungs, he spoke of a combination of condensation and rarefaction, with the blood flow itself determined by the valves. Here Descartes attempted to convert Harvey's vitalistic system into a fully mechanical one.

Fully as interesting was the reaction of Johannes Walaeus (1604–1649), who defended the circulation and praised the genius of Harvey in two letters published in 1641. Devising a series of new experiments on dogs with vascular ligatures, he helped to extend knowledge of the blood flow. This work was of sufficient weight to influence Harvey himself. Yet the reputation of Harvey was severely damaged in the eyes of Walaeus when he was told that the actual discoverer had been the Venetian statesman, Paolo Sarpi (1552–1623). After an extended investigation, Walaeus became convinced (1645) that Sarpi, basing his work on that of other authors going back to antiquity, had indeed discovered that circulation. Sarpi then, he wrote, had taught Harvey, who had proceeded to spread the doctrine under his own name.

In the two decades after the publication of the *De motu cordis* Harvey did suffer considerable abuse from James Primerose (d. 1659), Jean Riolan (1580–1657), Gui Patin (1601–1672), and a number of other authors, but it is also true that Harvey's work did not seem to everyone to be fully satisfactory. There still remained questions regarding the connection of the venous and the arterial systems, the different appearance of the two bloods, the role of the lungs, and the part played by the air in the formation of the arterial blood. Harvey himself had not been able to answer all these questions, but he did live long enough to receive the praise of most of his contemporaries. Shortly after his death the anastomoses between the arteries and the veins were first discovered by Marcello Malpighi (1628–1694) as he studied the blood flow in the lung of a frog with a microscope in 1659. Fourteen years later Antoni Leeuwenhoek (1632–1723) confirmed this observation with higher-powered lenses. The problem of the respiration was investigated by members of the early scientific academies, especially the Royal Society of London, and ill-fated experiments on transfusion seemed possible because of the new understanding of the blood flow. As to medical practice, the decline of the humoral theory resulted in a new interest in the chemical components of the blood. Numerous experiments were to be made in the late seventeenth century on the curing of disease through the infusion of chemicals.

But even late in the century the discovery continued to be of great interest to those who sought a mystical explanation of nature. Johann Rudolph Glauber, a late Paracelsian who contributed much to the development of large-scale chemical equipment, argued that the circulation of the blood had demonstrated conclusively the sympathetic connection of the macrocosm and the microcosm (1658). In England John Webster (1610–1682) pointed to the discovery of the mysteries of the blood made by Robert Fludd and William Harvey as subjects of importance that ought to be adopted as part of a needed curricular reform at Oxford and Cambridge (1654). And even the alchemist and early member of the Royal Society Elias Ashmole (1617–1692) referred to Harvey as a man "who deserves for his many eminent discoveries to have a statue erected rather of Gold than of marble" (1652).

The story of the discovery of the circulation of the blood may be easily cast in terms of the progressive growth of knowledge. It is possible to point to a whole series of discoveries in the period from Vesalius to Harvey. Nevertheless, these discoveries are set in a context of familiar themes. The anatomical renaissance of the sixteenth century was based upon several centuries of public anatomies at European universities. And here, as in botany, may be seen the power of the new union of art and observation. As for humanism, nowhere else was it to be more influential. The new Greek and Latin editions of Galen spurred a new interest in anatomical and physiological research. We have noted this in London with Thomas Linacre, in Paris with Guinter of Andernach, and in the whole series of Paduan teachers who were inspired by the work of the second-century Greek physician.

The new anatomy and the whole background to the discovery of the circulation is primarily associated with the University of Padua. Here there was a strong emphasis on the work of Aristotle and Galen, and we find their influence on the main figures in this development. They all approved of the correction of errors in the ancient texts, an exercise considered both right and proper, but there was no thought of overturning or replacing the ancients. Among these authorities only Guinter was to reflect favorably on the chemical remedies of the Paracelsians, but he did this without wishing to reject ancient medical theory. As for Harvey, he thought of himself as both an Aristotelian and Galenist until the end of his life.

And yet Renaissance anatomy and physiology is even more complex. When seen in context, the motives of many of these scholars differed drastically from that of modern scientists. Thus Servetus, trained as an anatomist, was prompted to set down his views on the blood flow because of his theological speculations. For Fludd the situation was similar. The understanding of nature was no less than an understanding of theology. It was on these terms that Fludd became the first to support Harvey in print. There were some who were to reject the Aristotelian Harvey because his views conflicted with the ancients whereas others were to support him, but only after stripping away the basic vitalism of his views.

In the end one is faced with the seeming paradox that one of the most impressive achievements of the Scientific Revolution was accomplished by a professed Aristotelian and that his work appealed first to mystical Hermeticists. And yet for all of this, the achievement was a great one and soon recognized as such. Others had known that the blood leaves the heart in the great arteries. But the earlier explanations had all involved a vast irrigation system with the purpose of building tissue. And if some had spoken earlier of a mystical circulation of the blood, Harvey now referred to real experiments and offered an irrefutable quantitative argument. It has been suggested that Harvey's work was the first adequate explanation of a bodily process, the starting point on the road to modern physiology. Certainly there was a change in attitude toward living processes after this time. Earlier references to indefinable innate heat, pneumatic force, animal spirits, and internal *archei* were to be replaced by a new search for simpler physical conceptions.

CHAPTER V

A New World System

The debates we have already traced in medicine, botany, and physiology in the late sixteenth and the early seventeenth centuries find a striking parallel in the fields of astronomy and physics. Once again we see the impact of humanistic research resulting in a significant restatement (Vesalius, 1543; Copernicus, 1543). And again, an extended period of assimilation and debate that was to result in new discoveries and interpretations essential for further development (Harvey, 1628; Kepler and Galileo, 1609–1632).

The problems related to an earth in motion involved not only a restructuring of the heavens, but also the development of a new physics of motion – the latter, a goal not fully realized until the publication of Isaac Newton's *Principia mathematica* in 1687. The full story of this latter development thus belongs properly to another volume in this series. But the acceptance of the Copernican system involves factors that go beyond the mechanics of motion and cosmology. Here again mysticism was influential in the motivation of key figures; no less important were theological problems that caused deep divisions.

Antiquity and Renaissance Astronomy

As in all other fields, Renaissance astronomy was based upon the work of the ancients. Two cosmological systems dominated this literature. The first, that of Eudoxus, Callippus, and Aristotle, employed a series of concentric spheres to account for the diurnal rotation of the stars and the motions of the sun, moon, and planets (Figure 5.1). To explain the variety of motions observed, the planets were assigned up to four spheres each. With a sufficient number properly arranged it was possible to account for movements as complex as the precession of the equinoxes and the retrogression of

74

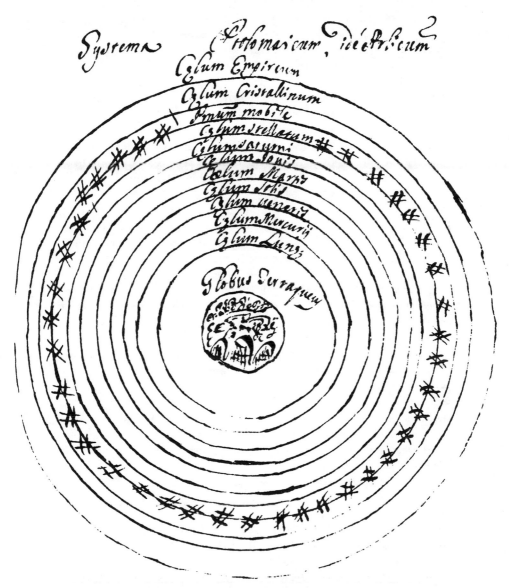

Figure 5.1. An illustration of the concentric spheres from a mid-eighteenth-century manuscript of an introduction to natural philosophy. Note the central earth surrounded by the spheres of the moon, the planets, and the fixed stars. From the collection of the author.

the planets against the background of the stars. This system, popular though it was, did not explain the fact that on occasion the sun, moon, and planets seemed periodically to be at different distances from the earth because their apparent size or brightness varied.

To solve this problem and to eliminate inaccuracies in the Aristotelian cosmology, Alexandrian astronomers of the third and second centuries B.C. (especially Apollonius of Perga and Hipparchus) recast the data into a new system. Their system was in turn revised and expanded into the detailed and truly mathematical system of Claudius Ptolemy in the second century A.D. It was this Ptolemaic system, as described in his *Almagest*, which commanded the allegiance of most astronomers until the seventeenth century.

Ptolemaic astronomy maintained the earlier spheres, but added a variety of circles (thus maintaining "perfect" motion in the heavens) to explain in greater detail the observed phenomena. In the simplest case a planet might be located on a great or deferent circle — if it appeared to move around the earth with perfect circularity. However, such perfection — other than for the stars — did not exist. As a result a number of other circles were postulated. The epicycle circle, with its center located on the circumference of the deferent, revolved while progressing with the movement of the deferent. This dual motion accounted for both apparent variations in distance and planetary retrogressions (Figures 5.2 and 5.3). Other irregularities led Ptolemy to place the earth some distance from the sun, and to use eccentric (off-centered) circles and equant circles. The latter accounted for apparent changes in planetary speed. In this case equal angles (originating from a point not at the center of the circle) were traced out in equal periods of time. Combinations of all such devices produced a sophisticated (if not perfect) astronomical system that both accounted for and predicted with a fair degree of accuracy the motions of the heavens.

But if Ptolemy and Aristotle overshadowed others in antiquity, their cosmological systems did not represent the full spectrum of ancient astronomical thought. In the fifth century B.C. some followers of Pythagoras (especially Philolaus) had suggested a universe in whose center was a central fire about which circled all the other bodies, including the earth, the sun, and a counter-earth. Later Aristarchus of Samos (third century B.C.) argued that everything would appear the same if the earth rotated daily on its axis and also revolved annually around the sun. However, this "Copernican" scheme was not worked out mathematically and only his calculation of the distances to and sizes of the sun and moon survived. More influential was the work of Heraclides of Pontus (fourth

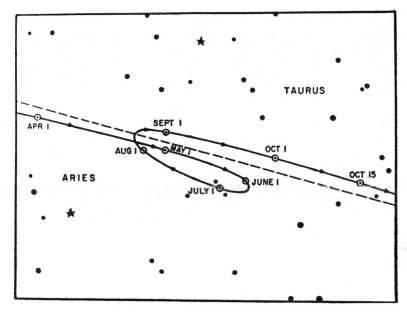

Figure 5.2. An example of the apparent retrogression of Mars viewed against the background of "fixed" stars. From Thomas S. Kuhn, *The Copernican Revolution* (Cambridge, Mass.: Harvard University Press, 1957), p. 48. Copyright © 1957 by the President and Fellows of Harvard College.

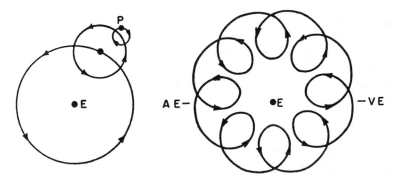

Figure 5.3. In the Ptolemaic system retrogression was explained in terms of a planet moving on an epicyclic circle attached to a larger moving deferent circle (left). The multiple rotations produced a series of loops or seemingly backward movements when viewed against the stars. From Thomas S. Kuhn, *The Copernican Revolution* (Cambridge, Mass.: Harvard University Press, 1957), p. 68. Copyright © 1957 by the President and Fellows of Harvard College.

century B.C.), who suggested that the diurnal revolution of the earth best accounted for the motion of the stars, and that the fact that Mercury and Venus were never seen far from the sun indicated that these planets circled the sun. Martianus Capella and Macrobius (both fifth century A.D.) repeated the views of Heraclides in late antiquity in texts that survived the barbarian onslaught and were used until the recovery of more detailed scientific works in the twelfth and the thirteenth centuries.

Islamic astronomers produced their own commentaries and revisions of Ptolemy and these — plus the original texts — were transmitted to Western Europe in the twelfth century. Introductory astronomical treatises [the most popular was that of John of Holywood (Johannes Sacrobosco) fl. 1230] were written, and translators prepared Latin versions of the cosmological works of both Ptolemy and Aristotle. Though later criticized as inaccurate, these translations remained influential down to the sixteenth century.

Aristotle was condemned on theological grounds in the course of the thirteenth century because his books on natural philosophy presented judgments that conflicted with Christian dogma. It was pointed out that the literal acceptance of his work would lead to the denial of the Creation by God, the truth of the Eucharist, the possibility of miracles, and the immortality of the soul. Once rejected as dogmatic texts, the Aristotelian corpus was opened to debate. In fact, a number of questions discussed in the late thirteenth and the fourteenth centuries did relate to astronomy and cosmology, such as the possible plurality of worlds and the motion of the earth. Such subjects were to be of special interest to Nicolaus Cusanus (1401–1464), who wrote of an unbounded (if not infinite) universe with all of its parts in motion. And if it is difficult to determine his exact understanding of earthly motion, there is little doubt that he rejected many of the views of contemporary astronomers.

The humanistic revival affected astronomy in many ways. Cusanus is one example of the new Platonic influence on fifteenth-century scholars. Marsilio Ficino is another. Ficino's mystical interests are reflected in his translation of the *Corpus hermeticum* and in his rhapsody to the sun (from the *De sole*), in which he followed the lead of earlier Hermetic texts:

"Nothing reveals the nature of the Good [which is God] more fully than the [light of the sun]. First, light is the most brilliant and clearest of sensible objects. Second, there is nothing which spreads out so easily, broadly, or rapidly as light. Third, like a caress, it penetrates all things harmlessly and most gently. Fourth, the heat which accompanies it fosters and

nourishes all things and is the universal generator and mover . . . Similarly
the Good is itself spread everywhere, and it soothes and entices all things.
It does not work by compulsion, but through the love which accompanies
it, like heat [which accompanies light]. This love allures all objects so that
they freely embrace the Good . . . Perhaps light is itself the celestial
spirit's sense of sight, or its act of seeing, operating from a distance, linking all things to heaven, yet never leaving heaven nor mixing with external
things . . . Just look at the skies, I pray you, citizens of heavenly father-
land . . . The sun can signify God to you, and who shall dare to say the sun
is false."

Ficino went on to note that the sun had been the first thing to be created
and that it was positioned in the center of the heavens.

Far different was the work of Georg Peuerbach of the University of
Vienna, whose *Theoricae novae planetarum* (published c. 1473) described in
technical terms an improved planetary system based on Ptolemy. Convinced that a better text of Ptolemy was required, he planned a trip to Italy
with his student and assistant Johannes Müller (Regiomontanus). At his
death Peuerbach had finished the first six books of an *Epitome* of the *Almagest* and Regiomontanus was to complete this work, which was first published twenty years after his own death (Figure 5.4). Publication of the
complete *Almagest* was to wait until 1515, when it appeared in Gerard of
Cremona's thirteenth-century version. A new translation from the Greek
was not to appear in print until 1528.

Copernicus and a Stationary Sun

We have already noted that the first phase of the Scientific Revolution involved a return to and a study of the ancient sources. For some this was a return to Aristotle and for others it was a return to Galen. A third group
sought truth in the search for the divine knowledge known to Adam (the
prisca theologia) through the study of the Hermetic texts. This Hermetic influence is evident in Renaissance cosmology in the writings of Ficino,
whereas a revived Aristotelian system of concentric spheres was described
by Girolamo Fracastoro (c. 1478–1553) in 1538. For him the previously
difficult problem of explaining varying distance was solved simply by assuming that the crystalline spheres were of variable transparency, thus giving the earthly observer the illusion of changing distance. But although
Aristotle and Hermes were influential, there is no doubt that Ptolemy
reigned supreme among humanist astronomers. And it was to Ptolemy that
Copernicus was most indebted.

Figure 5.4. Ptolemy and Regiomontanus – the frontispiece to the latter's *Epitome* of the *Almagest* (1496). Courtesy of the Newberry Library, Chicago.

Copernicus was born in Torun, Poland, in 1473, the probable date of the first edition of Peuerbach's astronomical text. At eighteen he matriculated at the University of Cracow, where he began his extensive collection of astronomical and mathematical books. In 1496 — the year of the first edition of the *Epitome* of the *Almagest* prepared by Regiomontanus — he set off for Bologna to study canon law. After a brief visit home (1501) he returned to study at Padua and took his degree of doctor of canon law at Ferrara in 1503. Both at Bologna and at Padua Copernicus had been in contact with learned astronomers. In Bologna he had known Domenico Maria da Novara (1454–1504) and in Padua there was Girolamo Fracastoro. The latter was a philosopher and physician as well as an astronomer and it was to medicine that Copernicus turned after taking his law degree. In later years he was to make use of his medical skills as part of his regular duties in Poland.

He returned permanently to his homeland in 1506, where he took part in the governing of the small state of Ermland, frequently participating in medical and economic decisions. In Italy Copernicus had learned Greek and one could list him as one of the minor literary humanists inasmuch as he published his translation of the poems of the seventh-century Byzantine author, Theophylactus Simocatta, in 1509. Although this was hardly a major event in humanistic circles, the volume is of interest for a prefatory poem that praises the translator for his astronomical pursuits. Even then Copernicus was becoming known as an astronomer; when calendrical reform was being considered in 1514 he was invited (and declined) to come to Rome to participate in the deliberations.

Although he had published nothing other than this short translation, many of his friends in both Poland and Italy knew of his interests. A few of the observations published in his *De revolutionibus orbium coelestium* (1543) date from his student days in Italy. But his conception of a heliocentric universe (heliostatic is a better word since Copernicus did not place the sun exactly at the center) was already fully worked out by about 1512 in the manuscript that is commonly called the *Commentariolus*. Here he outlined his theory and sketched some of its consequences — advising the reader that he was engaged in a larger study of the subject.

The idea of an earth in motion seemed to be contrary to common sense and it had presented innumerable difficulties to the ancients. Ptolemy had noted that if the earth moved, all objects not anchored to its surface would be left behind. A special case was the dropping of stones from appreciable heights. If the earth moved, the stone should be carried miles away in a few seconds of fall. The fact that the fall was observed to be straight seemed conclusive to those who maintained that the earth was at rest. Copernicus was to argue that the air is carried around with the earth (and thus the falling stone would be carried with it), but this argument seemed weak to many of his contemporaries.

Why should he have turned to a heliostatic system at all? It has been argued that had he not done so he would have had to accept the intersection of the spheres of Mars and the sun during the course of their revolutions. This might well have been unacceptable for one who still held to the existence of crystalline spheres. But there are other, perhaps less rational, considerations that cannot be ignored. In the *De revolutionibus* Copernicus asserted that his work merely revived the Pythagorean doctrines of antiquity. But, after relating the order of the planets, he concluded that

"in the center of all resides the Sun. Who, indeed, in this most magnificent temple would put the light in another, or in a better place than that one wherefrom it could at the same time illuminate the whole of it? Therefore it is not improperly that some people call it the lamp of the world, others its mind, others its ruler. Trismegistus [calls it] the visible God, Sophocles' Electra, the All-Seeing. Thus, assuredly, as residing in the royal see the Sun governs the surrounding family of the stars."

Whatever the reason, Copernicus had settled upon his system by his fortieth year if not earlier. Why then did he wait another three decades before publishing? In the preface to the Pope in the *De revolutionibus* he remarked that his reluctance came from a fear of the reaction of the ignorant. Astronomy was a subject for mathematicians, not the mob. There is, indeed, little indication of his desire to publish until a Lutheran scholar from the University of Wittenberg, Georg Joachim Rheticus (1514–1574), came to Ermland in 1539 with one purpose in mind, to learn more of this new theory of which he had until then heard little more than rumors. Copernicus, generous of his time, made known the fruits of his research of the past forty years and within twelve months Rheticus had prepared a brief manuscript describing the Copernican system. This *Narratio prima* reached a relatively wide audience and went through two editions (1540, 1541). Encouraged by this response, Rheticus urged Copernicus to publish his full work.

Relying on the promise of Rheticus to see the book through the press, Copernicus entrusted the manuscript to him. In fact Rheticus did not live up to his promise, and the published volume — which reached Copernicus just prior to his death — included an unsigned preface by Andreas Osiander (1498–1552), a Lutheran clergyman, who suggested that the system described in the book was primarily a mathematical device to facilitate astro-

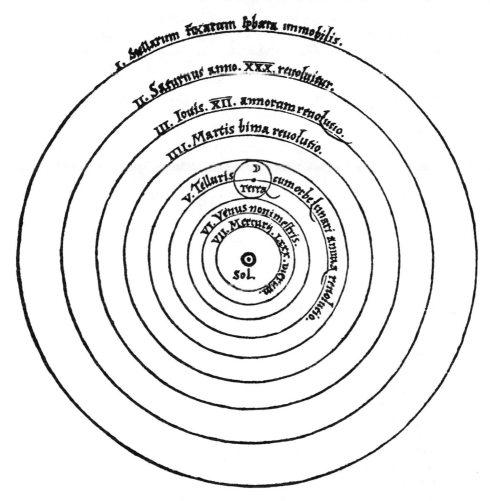

Figure 5.5. The Copernican system of the world, with a central sun surrounded by the orbits of the planets, including the earth with its lunar satellite. From Copernicus, *De revolutionibus orbium coelestium* (1543), sig. ci^v.

nomical calculations. This had hardly been the intent of Copernicus, but the fact that he was not the author of the preface was widely known later in the century.

The basic concepts of Copernicus are found in the first book of the *De revolutionibus* (Figure 5.5). Here the order of the planets is outlined and here too we are quickly made aware of the strong influence of Ptolemy. Indeed,

Copernicus is not known for his own observations, which for the most part were few in number — and certainly less precise — than those of some of his predecessors. Nor did Copernicus greatly simplify the older astronomy (Figure 5.6). He still accepted the deferent and epicyclic circles of Ptolemy, and he found that he could not place the sun precisely in the center any more than previous astronomers could place the earth in that place. And if one device, the equant circle, was eliminated as being physically meaningless, little else was totally discarded.

In short, the Ptolemaic system was recast. The sun now was at rest near (but not actually at) the mathematical center of the universe and it was surrounded by the planets (of which the earth was considered one, with its attendant moon on an epicycle) embedded in their crystalline spheres. The system included the time-honored sphere of fixed stars.

For Copernicus this system was simpler and more harmonious than earlier ones — and, as he noted, it gave a more appropriate place to the majestic sun. But if the Copernican system retained much of the complexity of the Ptolemaic universe, he simplified it to some extent. Not only were the equant circles eliminated, but the epicycles that accounted for the retrograde motion of the planets were now found to be unnecessary as well (Figure 5.7). The backward loop of the planets against the backdrop of fixed stars could now be accounted for as the result of the relative positions and speeds of the earth and the observed planets. The Copernican system also proved useful in determining the relative distances of the planets from the sun by simple trigonometric methods (Figure 5.8).

Stellar Parallax and the Size of the Universe

If the new explanation of retrogression was a triumph of the Copernican theory, other problems remained to trouble the astronomers of the late sixteenth century. The physics of an earth in motion was not to be solved until the next century, but the question of stellar parallax was to engage the attention of many sixteenth-century astronomers. If the earth does revolve around the sun annually, it was argued, the observer should experience a measurable shift in his view of any given star — at least if the universe was of the order of magnitude suggested by the ancients. But no such angular variation could be detected at six-month intervals. As a result it seemed to many that the earth must be at rest, and Copernicans had to argue that the universe was vastly larger than astronomers had suggested earlier (Figure 5.9). Acceptance of Copernicus also involved taking a stand on the size of the universe.

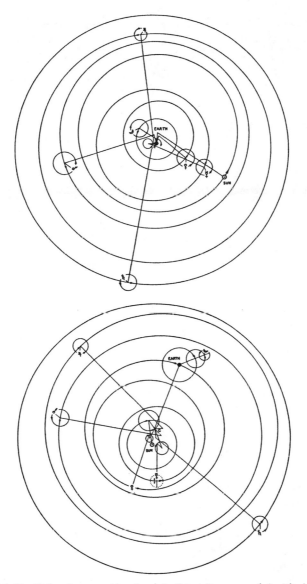

Figure 5.6. The Ptolemaic system (above) and the Copernican system (below) both made use of a variety of circular devices, so that the two were of the same order of complexity. From the diagrams of William D. Stahlman reproduced in Galileo Galilei, *Dialogue on the Great World Systems,* trans. T. Salusbury, ed., corrected, and with an introduction by G. de Santillana (Chicago: University of Chicago Press, 1953), pp. xvi–xvii. Copyright 1953 by The University of Chicago. All rights reserved. Published 1953. Composed and printed by The University of Chicago Press, Chicago, Illinois, U.S.A.

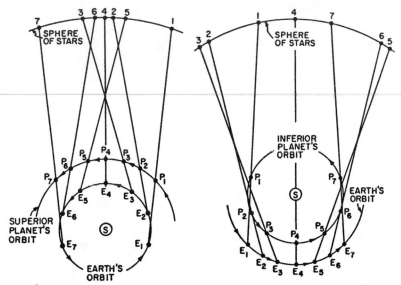

Figure 5.7. Retrograde motion in the Copernican system is explained in terms of the motions of the earth and the other planet being viewed. Once again — when viewed against the stars — there are apparent reversals in the planetary paths. From Thomas S. Kuhn, *The Copernican Revolution* (Cambridge, Mass.: Harvard University Press, 1957), p. 165. Copyright © 1957 by the President and Fellows of Harvard College.

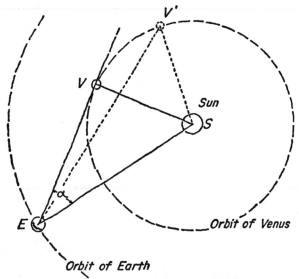

Figure 5.8. The Copernican system made possible the actual determination of the relative distances of the planets. Here the maximum value of the angle alpha can only occur when the line of sight from earth to Venus is tangent to the sun's orbit. Knowing the three angles it is not difficult to determine the orbital radius of Venus (or any other planet) in comparison to that of the earth. From I. Bernard Cohen, *The Birth of a New Physics.* Copyright © 1960 by Educational Services, Inc. Reprinted by permission of Doubleday & Company, Inc.

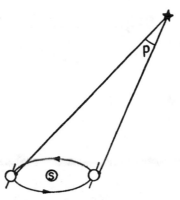

Figure 5.9. In the sixteenth century it was generally not thought that stars were an incredibly great distance from the earth. Therefore, if this were a sun-centered system the different positions of the earth at six-month intervals would require that the stars be seen at different angles of elevation. This annual angle of parallax was not actually determined until 1838. From Thomas S. Kuhn, *The Copernican Revolution* (Cambridge, Mass.: Harvard University Press, 1957), p. 162. Copyright © 1957 by the President and Fellows of Harvard College.

There was a background to this discussion. Cusanus had described an infinitely extended universe, and a similar system was proposed by Giordano Bruno (c. 1548–1600), who wrote of an infinite and decentralized world. In this way "is the excellence of God magnified and the greatness of his kingdom made manifest; he is glorified not in one, but in countless suns; not in a single earth, but in a thousand, I say, in an infinity of worlds." Bruno felt that he had elevated the earth to a new level, to that of the stars. But, he added, our earth does revolve around our sun and in a similar fashion an infinite number of earths in an infinite number of solar systems do the same.

Bruno's daring views were set in a context of neo-Platonic and Hermetic mysticism. And if his views were mixed with equally controversial theological speculations that were to lead to his death at the stake in Rome, his dismissal of the fixed sphere of stars was accepted with far less adverse reaction by others. In England Thomas Digges (c. 1543–1595) paraphrased the first book of the *De revolutionibus,* which he appended to a new edition in 1576 of a perpetual almanac (Figure 5.10). This is the most important sixteenth-century description of the new cosmological system in England, and it is important also because Digges eliminated the fixed sphere of stars, noting that the "orbe of starres" was "fixed infinitely up" and "extendeth hit selfe in altitude sphericallye." Therefore immovable is this "pallace of foelicitye garnished with perpetuall shininge glourious lightes," which are

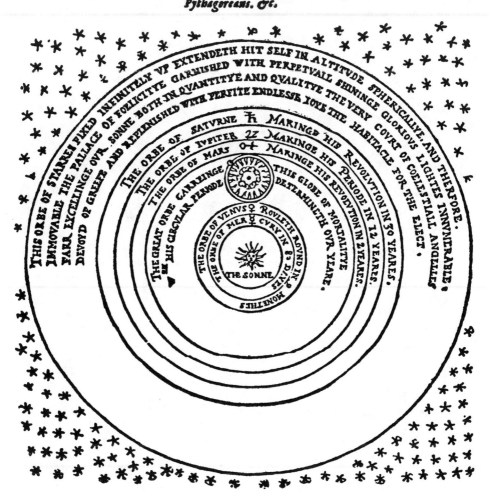

Figure 5.10. Thomas Digges's infinite Copernican universe (1576). From Francis R. Johnson, *Astronomical Thought in Renaissance England: A Study of the English Scientific Writings from 1500 to 1645*, p. 166. Copyright 1937, The Johns Hopkins University Press.

innumerable and "farr excellinge our sonne both in quantitye and quali-
tye." This is "the very court of coelestiall angells devoyd of greefe and
replenished with perfite endlesse ioye the habitacle for the elect."

Digges's diagram, in turn, may have influenced his compatriot, William
Gilbert (1540–1603), whose *De magnete* (1600) remains a classic of experimental method. Like so many other sixteenth-century authors, Gilbert
went far beyond the limits of science as we understand it. For him the
simple effects of the lodestone could clearly be amplified to permit an interpretation of the earth itself. In his description the earth was a magnet
and magnetism was best understood in terms of an animistic force. Unwilling to accept the entire Copernican system, Gilbert did subscribe to a diurnal rotation of the earth because he did not believe that the heavens could
revolve completely in a single day. Similarly, he rejected the ancient stellar
sphere, and in the posthumous *De mundo sublunari* (not published until
1651), he pictured the stars scattered throughout the infinite heavens in a
fashion similar to that of Digges (Figure 5.11).

But if the problem of parallax was being solved by many through the acceptance of an infinite – or at least a greatly expanded – universe, there
were others who sought to maintain a geocentric universe that might accommodate the increasingly more accurate astronomical observations.
Chief among these astronomers was the Dane, Tycho Brahe (1546–1601).
The son of a nobleman, he had been given a good education at the universities of northern Europe and seemed destined for a life in politics. But, attracted first by chemistry, he equipped a laboratory, only to be distracted
by the appearance of a "new" star (actually a supernova) in 1572. The
supernova was of major importance to all European astronomers because its
appearance clearly signified a change in the heavens. Traditional astronomers were quick to argue that this event must have occurred in the
imperfect sublunary regions because change could not occur in the higher
reaches of our world. But if the new star actually existed in the lower
regions (and was relatively close to the earth), it must be possible to detect
its parallax. Tycho, a brilliant and systematic observer, sought to determine this parallax but could find none. The new star must then exist at a
great distance from the earth. Thus, contrary to accepted belief, change
must be possible in the supralunary regions.

Of no less importance for cosmology was the series of comets Tycho observed between 1577 and 1596. In none of these cases did he observe
parallax, once again calling into question the doctrine of the immutability
of the heavens. Even more difficult for traditional astronomy was the fact
that these observations required the acceptance of the motion of these

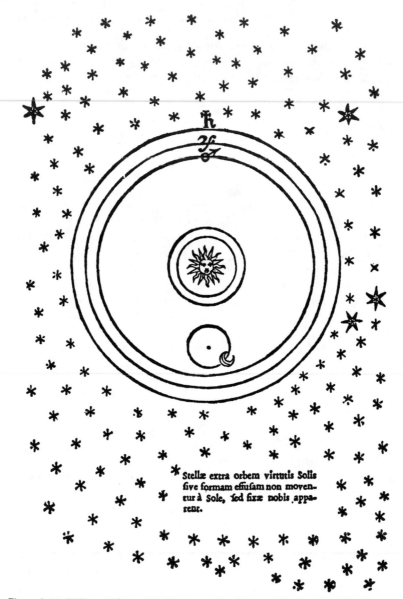

Stellæ extra orbem virtutis Solis
five formam effufam non moven-
tur à Sole, fed fixæ nobis apparent.

Figure 5.11. William Gilbert – like Digges – rejected the traditional sphere of fixed stars.
From his *De mundo sublunari philosophia nova*, published posthumously at Amsterdam (1651).
Courtesy of the Newberry Library, Chicago.

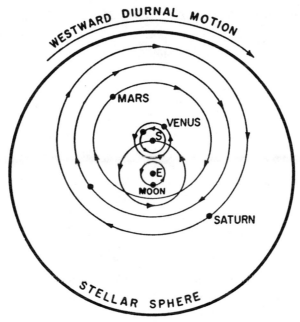

Figure 5.12. The Tychonic system of the world, showing a central earth being circled by the moon and the sun. The sun, in turn, has the orbits of the other planets surrounding it. From Thomas S. Kuhn, *The Copernican Revolution* (Cambridge, Mass.: Harvard University Press, 1957), p. 202. Copyright © 1957 by the President and Fellows of Harvard College.

comets through an area that was formerly thought to be occupied by crystalline spheres. It became therefore far more difficult to accept these spheres as physical realities.

But if his observations contributed to the weakening of ancient cosmology, Tycho himself found it difficult to accept the Copernican theory. Above all, the absence of stellar parallax required far too great a distance between the planetary orbits and the stars for his approval. He therefore introduced a compromise that maintained the stationary earth with its lunar satellite but required the circular motion of the sun about the earth — and the circular motion of all the other planets around the sun (Figure 5.12). The sphere of fixed stars was thus maintained at what seemed to be a reasonable distance from the earth, sun, and planets. Mathematically the system was similar to that of Copernicus, and Tycho maintained the various Ptolemaic circular devices to ensure the accuracy of the system. However, Tycho had managed to avoid the necessity of coping with the problem of a new physics of motion — a problem that was becoming increasingly troublesome by the end of the century.

Solving the Mysteries of the Planetary Orbits

Tycho Brahe was recognized by all as the leading observational astronomer in Europe. Not only did he make regular observations of the planets and prepare the way for a more accurate set of star tables than had ever existed, but he also designed observational equipment on a scale surpassing in size any from earlier times. This equipment was installed in a castle-observatory built at Hveen by the Danish king. Here a staff of assistants made nightly observations of the stars, and "terrestrial astronomy" (chemistry) was studied by others in the extensive chemical laboratories on the lower levels.

Recognizing the preeminence of Tycho in his field, a young German sent a copy of his first work to him in 1596. This was Johannes Kepler, who was eventually to become Tycho's heir and greatest disciple. Kepler had become a Copernican early in life. Sent to study at Tübingen, he heard the lectures of Michael Maestlin (1550–1631) on astronomy. Although these lectures were based on Ptolemaic theory, Kepler later related that Maestlin had also discussed the work of Copernicus. The new astronomy appealed to the student; when he left to take a position as mathematician (and astrologer) at Graz (1594) he was already engaged in a work on Copernican astronomy.

Kepler's first book was the one that would be sent both to Tycho and to a yet undistinguished Italian professor of mathematics, Galileo Galilei. The work itself displayed Kepler's great mathematical talent and his persistent interest in mystical relationships. Entitled *Mysterium cosmographicum* (1596), Kepler here gave his first answer to his quest for a universal mathematical order. Convinced that there must be some consistent interrelation of the planetary orbits, he calculated and recalculated their distances from the sun. His conclusion was that the planets all bear a distinct relationship to the five regular solids. Thus, he thought, the universe could be accurately depicted with the sun at the center and with the planetary spheres of Mercury, Venus, Earth, Mars, Saturn, and Jupiter separated from one another by first an octahedron, then an icosahedron, a dodecahedron, a tetrahedron, and a cube (Figure 5.13). This result clearly reflects the contemporary interest in neo-Pythagorean numerical mysticism, and to Kepler the discovery was fundamental because it showed the mathematical order of the universe.

Tycho was favorably disposed toward Kepler and his book — so much so that he offered him a position at Hveen. Kepler declined, preferring to remain at Graz, but over the next few years the religious climate became increasingly unhealthy for a Protestant and he finally decided to leave Graz

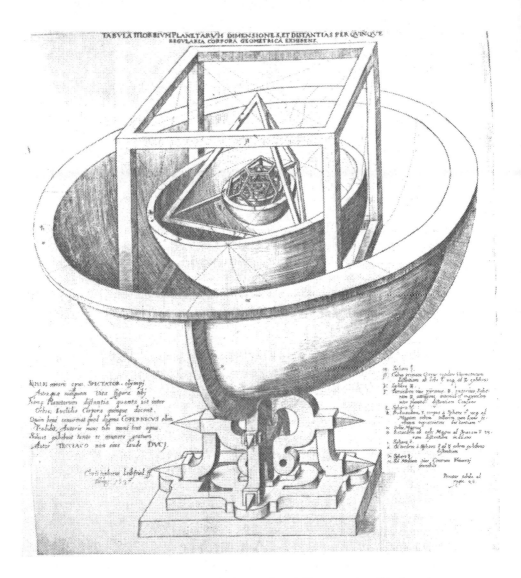

Figure 5.13. Kepler's model of the planetary orbits bounded by the regular solids. From the *Mysterium Cosmographicum* (1596). Courtesy of The Joseph Regenstein Library, The University of Chicago.

in 1600. By this time Tycho had moved from Denmark to the court of the Emperor Rudolf II (reigned 1576–1612) at Prague. Here indeed was a center for European intellectuals. Rudolph had fostered all the sciences (including alchemy and astrology) since his accession. John Dee had been a long-time resident at the court and numerous alchemists and astrologers were on hand at all times. Although Tycho was appointed Imperial Mathematician it would seem likely that he was welcomed no less as a chemist than as an astronomer. His position permitted him the privilege of assistants, and it was just then that Kepler wrote to inquire whether he might still have use for his talents. Upon confirmation that Tycho's earlier offer still held, Kepler set out for the imperial court.

In Prague Kepler had access to Tycho's extensive observations on the planets, and before the latter's death in 1601 he was studying the data on the orbit of Mars in the hope that this information might be reduced to a regular mathematical rule. His first attempts involved the employment of normal Ptolemaic devices such as epicyclic and eccentric circles. The results, however, were not as good as he expected from Tycho's accurate data.

Abandoning this approach, Kepler turned next to the problem of the cause of planetary motion. The recent studies of the comets had made it necessary to abandon the crystalline spheres and so there had to be some other explanation. Influenced by Gilbert's magnetic forces, Kepler postulated a solar *anima motrix* similar to Gilbert's magnetism, a motive soul emanating from the sun that moved the planets in the course of its own axial revolution. This force, he believed, followed an inverse-square law in regard to distance – but only in the plane of the ecliptic. The result of these speculations is to be found in what is now commonly termed Kepler's second law of planetary motion, which states that a line from the sun to a planet sweeps out equal areas in equal times. Proceeding next to a mathematical study of the *anima motrix,* he concluded that the orbit of Mars was not a circle – a basic break with tradition inasmuch as circular motion had expressed the concept of perfection in the heavens. An extensive study of other possible curves resulted finally in the ellipse as the description of the orbital path (first law). These two conclusions, announced first in the *Astronomia nova* (1609), were based upon physical assumptions involving a vitalistic universe. However, the results held true and Kepler proceeded to apply them to the planets other than Mars.

Kepler's third law, which appeared in his *Harmonices mundi* in 1619, was also the result of his persistent search for the universal harmonies in nature. In modern terms this law states that the squares of the times of revolution of any two planets around the sun are proportional to the cubes of their mean distances from the sun. A brilliant discovery from our point of view, this relationship had a deeper meaning for Kepler. Literally seeking a mathematical expression of the world harmonies, he began this work with the study of the five regular solids and their harmonic ratios. From there he went on to musical harmonies and their relationship to the universe. The eighth book was to be devoted to a study of the four kinds of voice emitted by the planets (soprano, contralto, tenor, and bass), and the third law of planetary motion had been made part of a book on the expression of the clefs of the musical scale and the genera of major and minor consonances. In short, Kepler's laws of planetary motion were developed by a master mathematician, but if we are to judge them in their true context, they must be examined in relationship to Kepler's full world view.

Two years later Kepler prepared an *Epitome* of Copernican astronomy, in which he took the opportunity to reaffirm the results of his earlier *Mysterium cosmographicum*. His final years were devoted primarily to the completion of Tycho's planetary tables which appeared at last in 1627, three years prior to his own death.

The Physical Problem

Much of Kepler's most significant work was buried in the midst of his philosophical speculations, and it was not until the middle years of the seventeenth century that we see numbers of scholars referring to his laws as a basis of planetary theory. In the meantime a series of new developments — chiefly associated with the work of Galileo Galilei — hastened the acceptance of the Copernican theory.

Born and educated at Pisa, Galileo soon became interested in mathematics and astronomy. A strong influence on the young scholar was the work of Archimedes, whose mathematical expression of physical phenomena seemed to him to be far removed from the writings of Aristotle. He attacked the latter's works for their lack of mathematics and for their uncritical reliance on sensory experience. At least as a young scholar Galileo felt free to refer to the macrocosm—microcosm analogy as a true expression of a world with the sun as the king and heart of its planetary attendants. And Galileo, like Kepler, sought to mathematicize the whole universe, nature as well as supernature. Galileo became professor of mathematics at the University of Padua in 1592 but was not yet known as an astronomer when he was sent a copy of *Mysterium cosmographicum*. But whereas Tycho had read it in detail and had offered Kepler a position at Hveen, Galileo did little more than acknowledge receipt of the book and note that he, too, was convinced of the truth of the Copernican explanation of the world. There is nothing to indicate that Galileo was influenced in any way by Kepler, and in spite of his mathematical interpretation of local motion, he steadfastly adhered to circles in describing the movements of the sun, moon, and planets.

Although a number of early tracts by Galileo exist, including a lecture on the comet of 1604, he wrote nothing of major importance until 1610 when his *Sidereus nuncius,* or "Starry Messenger," a twenty-nine-page booklet, startled the intellectual community of Europe. It was written in Latin and gave the earliest published report of the telescopic observation of the heavens. The book gave powerful support to those who favored the Copernican system.

Although there is evidence to indicate that the telescope was fairly well known for a generation prior to the publication of the *Sidereus nuncius* — and even that some astronomical observations had been made earlier — there is no doubt that Galileo was the first to describe his discoveries in print. Here he discussed and pictured the lunar landscape (Figure 5.14). And even though his telescope was a weak one by our standards, it was accurate enough for him, knowing the size of the moon, to measure the shadows cast by lunar mountains and to calculate their height. He noted further that the telescope made visible a vast number of stars never seen before. And of special interest was his discovery of the moons of Jupiter, which he named the "Medicean" stars in honor of the Florentine ruling family. These circled that planet like a minature solar system. Further observations in the next few years were to reveal to Galileo the important fact that Venus exhibited phases like the moon. These could only occur if that planet circled the sun. The doctrine of the perfectibility of the heavens was further weakened by his disclosure that the sun had surface spots that rotated, thus indicating the axial motion of that body.

The *Sidereus nuncius* had its desired effect. His naming of the Jovian satellites in honor of Cosimo de' Medici had resulted in his appointment as Philosopher to the grand duke, thus making it possible to return to Florence. But he had also become a celebrity overnight, one who was compared with Columbus and Vespucci for his discovery of a new world. Kepler, writing of the new discoveries, rhapsodized: "O telescope, instrument of much knowledge, more precious than any scepter! Is not he who holds thee in his hand made king and lord of the works of God!"

Galileo's telescopic observations had shown the existence of a Jovian system of satellites similar to the Copernican planetary system and they had proved conclusively that Venus must revolve around the sun. These observations plus Galileo's open support of the heliocentric theory occurred at a

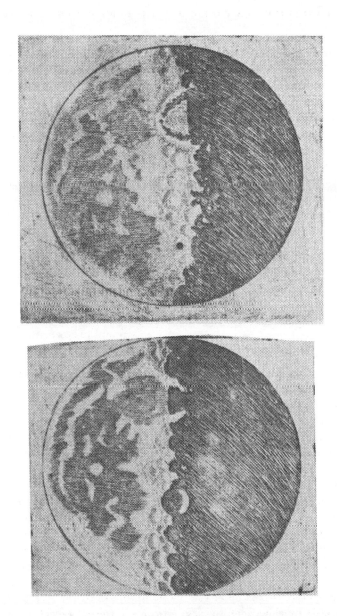

Figure 5.14. Galileo's illustrations of the moon. From the *Sidereus Nuncius* (1610). Courtesy of The Joseph Regenstein Library, The University of Chicago.

difficult period for the Roman Catholic Church and were to bring down upon him its wrath. Until this time the Church had remained silent, but now, in the midst of its own reform movement, it began to react to the dangerous theological implications of a sun-centered system of the universe. A century earlier Copernicus had been invited to participate in the planned calendrical reform – and in 1551 his mathematical methods of computation had been used as the basis for a new set of astronomical tables prepared by Erasmus Reinhold (1511–1553). Far less promising had been the immediate reaction from the Protestant camp. Martin Luther (1483–1546) referred to Copernicus as that fool who wished "to reverse the entire science of astronomy" (1539). He was seconded by Philip Melanchthon (1497–1560), who cited the Bible at length on behalf of the traditional world view (1549):

"The eyes are witnesses that the heavens revolve in the space of twenty-four hours. But certain men, either from the love of novelty, or to make a display of ingenuity, have concluded that the earth moves; and they maintain that neither the eighth sphere nor the sun revolves. . . Now, it is a want of honesty and decency to assert such notions publicly, and the example is pernicious. It is the part of a good mind to accept the truth as revealed by God and to acquiesce in it."

In 1616 the Holy Office branded the heliocentric theory as "foolish and absurd philosophically, and formally heretical, in as much as it expressly contradicts the doctrines of Holy Scripture in many places, both according to their literal meaning, and according to the common exposition and meaning of the holy Fathers and Doctors." A few weeks later *De Revolutionibus* was placed on the list of prohibited books and Galileo was warned not to defend the Copernican theory as a description of the actual physical constitution of the world.

Further work by Galileo on the theory of tides was to convince him that he now had proof of the earth's motion. Accordingly he planned and wrote his major work contrasting the Ptolemaic and Copernican systems, the *Dialogue on the Two Principal World Systems* (1632). Permission to publish was granted with the stipulation that he should discuss the Copernican system only as a hypothesis. This he did, but his discussion was far from impartial. The proponent of the Copernican system in this dialogue represented Galileo and thoroughly demolished the advocate of the old astronomy at every turn. And although at the close of the book Galileo dutifully reported that all he had said had been meant to be hypothetical, he had surely violated the warning given to him in 1616. It is thus hardly surprising that he should have been brought to trial by the Inquisition and obliged to abjure his belief in the condemned Copernican theses (1633). Confined for the rest of his life to his villa at Arcetri, he continued his work and his *Mathematical Discourses and Demonstrations Concerning Two New Sciences* was published in Holland in 1638.

These two final volumes were largely concerned with the problem of motion, which Galileo rightly understood to be inseparable from the cosmological systems he compared. A discussion of Galileo's views on motion need not be attempted here; it is enough to note that from the time of Copernicus it was realized that the problems of local motion on a moving earth would have to be explained. Galileo's own investigation of the laws of kinematics and dynamics did not reach the level of Newton in the *Principia,* but he did come close to the modern concept of inertia and he explained why a stone dropped from a tower falls at its foot, why a ball in the hand of a moving person falls back into the hand after having been thrown up vertically, and why two horsemen riding next to each other could throw balls across to each other and not have them fall many feet to the rear. If these explanations were correct, then one of the most telling arguments against the diurnal motion of the earth could be dismissed. Galileo's mathematical expression of the laws of motion was essential for the development of modern mechanics and was to be a basis for the work of Isaac Newton.

The Copernican system was not universally accepted by the time of Galileo's death in 1642. Most European authors still adhered to the Tychonic system, which certainly remained the safest choice in Roman Catholic countries. Nevertheless, the astronomical scene had radically altered since the time of Georg Peuerbach. His recognition of the need for a more accurate text of Ptolemy had resulted in the *Epitome* of 1496. And Copernicus's careful study of Ptolemy had resulted in turn in a recasting of Ptolemaic astronomy in heliostatic form. This, however, had brought a new set of problems for both astronomers and natural philosophers, the solution of which was to be the subject of debate for the century following the death of Copernicus. The search for measurable stellar parallax was to remain fruitless until the nineteenth century, but those inclined toward the Copernican system in the sixteenth century seem to have found little difficulty in accepting a greatly enlarged, and, for some an infinite, universe. Tycho maintained a geostatic system with a sphere of fixed stars, but even he employed a heliocentric system for planets other than the earth.

The power of mathematics was clearly demonstrated by the scientists discussed in this chapter. Copernicus gave credibility to his work by his mathematical treatment of Ptolemy's data, and he insisted that the entire subject of astronomy was one that belonged properly to mathematicians. Tycho immediately recognized Kepler's mathematical genius, and both Kepler and Galileo exhibited the power of their analyses through their mathematical treatments. And yet it is in Kepler that we see the best example of a Renaissance scientific paradox — the superb mathematician whose inspiration derived from his belief in the mystical harmonies of the universe. This heady mixture of mysticism and mathematics is far removed from modern science, but it formed an essential ingredient of its birth.

CHAPTER VI

New Methods and a New Science

The sixteenth century is one of paradox. It was a period of deep veneration for ancient authority, a veneration that stimulated some of the most renowned scholars of the period. Scientific humanism may be represented at its best by Peuerbach and Regiomontanus in astronomy and Linacre and Guinter of Andernach in medicine. The work of Copernicus and Vesalius simply cannot be understood outside a Ptolemaic or a Galenic context. And even a century later William Harvey was to think of himself as an Aristotelian and proclaim his debt to Galen. But for these great figures of the Scientific Revolution, respect and admiration of the ancients did not preclude their correction. This characteristic of humanism resulted in an ever-increasing volume of additions and alterations that was in time to swamp and overturn those very authorities the new work had been meant to uphold.

But the new mass of data had been fostered also from a less likely source. Renaissance humanism had resurrected the authority of Hermes Trismegistus no less than it had Ptolemy and Galen. With the *Corpus hermeticum* had come a new veneration for alchemy, natural magic, and astrology. And if the one strain of humanism had fostered a new study of mainstream Hellenistic science and medicine, the other emphasized the recovery of a *prisca theologia* presumed to have been known to man before the Fall. The Hermeticists had nothing but distrust for the work of Aristotle, Galen, and their followers. True science, they argued, was only to be found in the writings of those later magicians and alchemists who perceived the eternal truths underlying their endeavors. In the case of Paracelsus a true philosophy of nature could only be reached through the destruction of the authority of the ancients and its replacement with the divine knowledge to be obtained from a fresh — and largely chemically inspired — investigation

101

of God's created universe. In short, if the work of many Renaissance astronomers, mathematicians, and physicians was built upon the Hellenistic authors of the period from Aristotle to Ptolemy and Galen, there were others who saw the possibility of truth only in a complete overthrow of the science and medicine of the schools.

Regardless of individual appraisals of the value of the work of the ancients, it was becoming increasingly common in the late sixteenth century for scholars to think in terms of a new philosophy. As early as 1536 Petrus Ramus had defended the thesis that "everything which Aristotle states is false." In later life he was to devote his major effort to the development of a new and highly influential system of logic — a work that was aimed at the very foundation of scholasticism. And we have already noted Bernardino Telesio late in the century attacking medieval Aristotelianism in his academy at Cosenza. Rather than merely repeating Aristotle, he emphasized new studies of nature as a basis for the foundation of knowledge. We have also seen that William Gilbert's study of the magnet was for him the basis for an explanation of the world system. Gilbert was acutely aware of the novelty of his work, which "is almost a new thing unheard of before . . . Therefore we do not at all quote the ancients and the Greeks as our supporters." And even William Harvey, with his deep regard for both Aristotle and Galen, professed "both to learn and to teach anatomy, not from books but from dissections; not from the positions of philosophers, but from the fabric of nature."

But how was one to proceed? Should the scholar simply collect vast amounts of new facts indiscriminately or should there be a new plan and method of analysis for a new philosophy of nature? Harvey felt comfortable abstracting Aristotle's *Posterior Analytics* as a guide for his readers, but many others would have disagreed. The range of thought actually proposed at this time may be illustrated by the work of three men — Bacon and Descartes, who openly sought a "new philosophy," and Galileo, whose methodology is best exhibited through an actual example.

Francis Bacon

The work of Francis Bacon has long attracted the interest of historians. Lord Chancellor of England, he was the chief political architect of James I's program in Parliament until he was discovered taking bribes in 1623. He has been highly esteemed by literary critics for his *Essays* and histories, but his major effort was directed to the reformation of our knowledge of nature. Indeed, his unfinished "Great Instauration" was so influential in the second half of the century that it is safe to characterize much of the work carried out in the fledgling scientific societies and academies as "Baconian" in inspiration.

But whereas Bacon has been for years set forth as the standard-bearer of the inductive method in science, recent research has indicated his deep debt to unexpected sources. Well read in the literature of natural magic and alchemy, Bacon deplored the traditional secrecy of these subjects and argued that a true magician would make public his discoveries. But, he added, natural magic is not useless. Indeed, it is that "science which applies the knowledge of hidden forms to the production of wonderful operations; and by uniting (as they say) actives with passives displays the wonderful works of nature." How very similar this is to John Dee or to Paracelsus!

Further, Bacon was in agreement with many of his alchemical and Hermetic contemporaries in regard to the search and possible recovery of the pristine knowledge known to Adam. For this reason he studied in detail the myths of antiquity, seeing in them a link with the earliest human traditions. Thus he interpreted the story of Saturn allegorically in terms of Democritean atomism and he described the doomed quest of Orpheus for Eurydice as a symbol of the impotent haste of those who abandon experiment in their quest for knowledge. And in his hands the myth of Cupid (matter) became an analysis of the elemental systems of the pre-Socratics.

Bacon's belief in the Adamic wisdom was coupled with an equally recognizable Paracelsian trait, the blanket rejection of Aristotle. Convinced that a new era of history was at hand, he attacked the schools for their sterility and their concern for the maintenance of established texts rather than the much-needed search for progress seen so admirably in the mechanical arts. Aristotle was to blame for having attacked and rejected the work of earlier philosophers, thus breaking the last link with that pristine knowledge so necessary for mankind. Furthermore, he had subjected science to logic and had brought forth experiments only to support preconceived conclusions. And finally, his philosophy had been incorporated into religion and thus used to support the Roman Church. It was barely believable that this man's work was still so valued and that so little new had been discovered since his time.

What was to be done? The first step was to discard the accumulated Greek corpus along with its more recent commentaries, or, at the very least, to begin to examine these works without the blind reverence too often exhibited in the schools. Then scholars must begin to set up new catalogues of facts, observations, and experiments. This must all be done carefully, for only after the work was completed could true theories and natural laws be separated with relative ease.

But a definite plan would have to be followed in this endeavor. For Bacon the pure empiricists might be compared with ants, men who did no more than accumulate vast collections of facts. The philosophers were no better, and he likened them to spiders who, with their logic, spun complex webs from their own bodies. The true scientists, he explained, were best compared with bees, who extracted matter from flowers and then re-fashioned it into honey, useful to all. Bacon's new philosophy was to be ex-perimental, but his experiments were to be carefully chosen and always recorded in detail. Listing over one hundred and thirty important topics and processes for examination, he stressed the need for a great body of data that would be carefully classified. For each there would be a list of positive instances (where the phenomenon was present), a list of negative instances (where it was absent), and a list of degrees of comparison (where the phe-nomenon varied according to other factors). From such initial lists Bacon thought that knowledge might be obtained by excluding the improbable hypotheses and then proceeding to test the others.

Bacon's view of scientific method was essentially experimental, qualita-tive, and inductive in nature. Like the Paracelsians he distrusted mathe-matics. And if he stated that the investigation of nature was best conducted through the application of mathematics to physics, he also complained that it might be used to excess, and indeed he felt that the mathematicians were beginning — improperly — to dominate the subject.

He gave notice of his projected work in *The Advancement of Learning* (1605), and his "Great Instauration of Learning" was to include that work as a general introduction, a detailed analysis of scientific method (the *Novum Organum*, 1620), and a vast encyclopedia of craft lore and experi-mental data estimated by him to amount to six times the bulk of Pliny's natural history. The final sections were to include a discussion of previous and current scientific theory, plus the new natural philosophy as it emerged from the accumulated materials.

Needless to say, the entire project was far too much for him — or anyone else — to come close to completing. And if *The Advancement of Learning* and the *Novum Organum* were finished, many other sections were never at-tempted — or at best exist only through prefatory materials or other short sections. Yet the dream of a science emerging from masses of data was to inspire many seventeenth-century authors, who looked to Bacon as their guide. For many of these "Baconians" a posthumous work, the *Sylva syl-varum* (1627), was his greatest legacy. Here was to be found a mass of facts arranged in "centuries," mixing personal observations with notes from other varied sources. The result is strangely reminiscent of a Renaissance "book of secrets" in the natural-magic tradition. No other work shows so clearly the essentially unworkable nature of Baconian method. And yet this book went through at least fifteen English editions in the seventeenth century and moved no less an author than Robert Boyle to attempt its continuation.

One can say that Bacon's idea of a new science placed too little emphasis on mathematics and too much on experiment. He himself seemed unable to assess correctly the science of his own day. He questioned the value of the microscope and the telescope even though he frequently referred to Galileo's discoveries in the *Sidereus nuncius*. He criticized William Gilbert for having attempted to construct an entire philosophy on the basis of a single phenomenon, and he could not accept the Copernican system because he found no real evidence for the diurnal revolution of the earth. And if he frequently attacked the works of Paracelsus, in his own attempt to construct a cosmology he relied heavily on contemporary chemical theory. Thus the heavens were interpreted in terms of the sulfur-mercury theory and stellar motion was related to properties of the celestial fire. If, however, Bacon's scientific method is seen in terms of the full intellectual spectrum of his world, we see that he was influenced not only by the widespread discontent with scholastic method in the sciences, but also specifically by natural magic, alchemy, and the Paracelsian chemical philosophy.

René Descartes

Hardly less influential than Bacon was Descartes. Here again we meet with an attempt to establish a new, universal philosophy meant to replace that of the ancients. Educated at a newly founded Jesuit college, Descartes related later that he had "found himself embarrassed with so many doubts and errors that it seemed to me that the effort to instruct myself had no effect than the increasing of my own ignorance." Perhaps, he added, the whole body of the sciences need not be reformed, "but as regards all the opinions which up to this time I had embraced, I thought I could do no better than endeavour once and for all to sweep them away, so that they might later be replaced . . ."

In 1618 Descartes left France to enroll as an officer in the military academy of Prince Maurice of Nassau. The following year in Germany, while in deep meditation on November 10, 1619, he dreamed of a universal science of nature to which the key would be mathematics and the mathematical method. The account is reminiscent of the dreams so prominent in the contemporary alchemical literature. And in fact we do know that Descartes was then aware of the educational and scientific reforms being proposed by those neo-Paracelsian authors who wrote under the name of the "Rosicrucians." Further, on Descartes's return to Paris in 1623 he found that his friends feared that he had become a Rosicrucian while he had been away — an opinion that he found necessary to refute. This episode may only be accorded a footnote in most accounts of the work of Descartes, but it once more illustrates the difficulty faced by historians who seek absolutely to demarcate the "rational" from the "irrational" in the early seventeenth century.

In 1628 Descartes settled in Holland, where he devoted himself to his research. From there he maintained a steady correspondence with scholars everywhere, and especially with Father Marin Mersenne in Paris, whose monastic cell acted as a central clearing house for the European scientists of the period. By 1633 Descartes was ready to publish his *Le monde,* but on hearing of the condemnation of Galileo he suspended publication because of its Copernican nature. Descartes's major works were to appear only later. In 1637 he published his *Discourse on Method,* which served as an introduction to his longer studies on *Dioptrics* (including the lens, vision, and, the law of refraction), *Meteors* (including his study of the rainbow), and *Geometry* (including his development of analytical geometry). After that there appeared the *Meditationes de prima philosophia* (1641) and the *Principia philosophiae* (1644). Descartes was to return to France in 1647, only to be lured to Sweden two years later by Queen Christina. There he died in 1650.

Descartes shared with Bacon the desire to found a new philosophy free from older opinions. However, he went much further than Bacon in his disdain for tradition. For Descartes nothing less would suffice than to discard completely all past knowledge and to begin anew, accepting as axiomatic only God and the reality of one's own existence (*Cogito, ergo sum*). The Deity for him was known through the mind — indeed, the truth of God known in this fashion was far more evident than anything seen through the eyes. From this foundation Descartes was prepared to deduce the entire universe and its laws. It was his belief that each step in this mathematically inspired method would be as certain as the proofs in Euclidean geometry. It is no wonder that he was successful in his study of optics, the rainbow, and analytic geometry: These subjects are essentially mathematical and are best treated in this fashion.

In his cosmology Descartes proceeded from God to matter and motion.

He felt such confidence in his results that he was convinced that regardless of the number of separate universes that might have been created, they would necessarily have developed like our own. He was satisfied that his essentially deductive system had led him safely to a confirmation of the elements of matter, but when deduction resulted in a variety of possibilities, experiments were to be devised to make the actual choice.

Descartes's universe was "mechanical" and he rejected the vitalistic explanations so prevalent among his contemporaries. He postulated a constant quantity of motion in the universe; this was inherent in particles of three sizes that corresponded to the ancient elements of earth, air, and fire. The first, the largest, accounted for the chemical and physical properties of matter. The second, much smaller and more rapid in their motion, were to be found between the atoms of earth. Finally, the particles of fire, having a very violent motion, were to be found in whatever openings that might still exist. In this fashion all space was to be filled. As a result, Descartes (along with Aristotle) rejected the vacuum and the possibility of action at a distance. This was an attempt to explain all things in terms of vortices, whirlpools, of matter. Local aggregations formed the planets and the sun, and similar processes took place around the distant stars. The end result was a vast system of vortices that accounted for all the matter in the universe (Figure 6.1).

Descartes's mechanical philosophy when applied to man and biology stripped away the "vital" forces that had dominated previously. His own work was to play a significant role in the development of the iatrophysical school of the late seventeenth century. We have seen his approach earlier in his mechanical "correction" of the Harveian circulation. For him man was a union of a soul with a machinelike animal body, and he found it quite satisfactory to compare the workings of the body to the works of hydraulic engineering so frequently in evidence in the gardens of the rich in the early seventeenth century:

"one may very well liken the nerves of the animal machine I have described to the pipes of the machines of those fountains; its muscles and its tendons to the other different engines and springs that serve to move them; and its animal spirits, of which the heart is the source and the ventricles of the brain the reservoirs, to the water that moves these engines. Moreover, respiration and other similar functions which are usual and natural in the animal machine and which depend on the flow of the spirits are like the movements of a clock or of a mill, which the ordinary flow of water can make continuous."

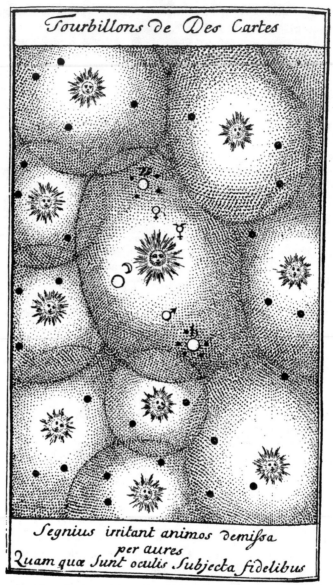

Figure 6.1. The Cartesian cosmology as pictured in an eighteenth-century text. Note the solar system in the center. From *Traité de l'opinion, ou mémoires pour servir à l'histoire de l'esprit humain* (Paris: Charles Oliver, 1733), vol. 3, plate 2. From the collection of the author.

External objects then triggered responses in the brain in a fashion analogous to the visitor entering a grotto who, stepping on plates in the ground, activated the figures on display. Thus, the body was viewed as a piece of machinery, its reflexes controlled by a vast tubular network in which valves opened and closed, allowing fluids to move in different directions and for different purposes.

Descartes was to have an influence on Continental science that was not to decline until the mid-eighteenth century. But if the defect of Bacon's new science was an overemphasis on experiment, that of Descartes failed as a result of its overemphasis on deduction. Truly effective only when applied to subjects that did in fact lend themselves to a mathematical treatment, few of his cosmological or biological speculations proved to have lasting value for the future development of science.

Galileo Galilei

The influence of Bacon and Descartes notwithstanding, both were wedded to methodologies that were badly flawed from the standpoint of the modern scientist. What was needed was more of a real interplay between the inductive and deductive processes. This approach is seen best in the work of Galileo — not in the form of a discussion of scientific method, but rather in the actual development of his subject. For our purpose the most useful volume to examine is the *Mathematical Discourses and Demonstrations Concerning Two New Sciences* (1638). There is no need to summarize its contents because it is not our intent to follow the physics of motion of the seventeenth century in detail. Nevertheless, Galileo's development of the problem of free fall presents an excellent example of his methodological procedure.

Beginning with the observation that in most studies of natural phenomena it is customary to seek their causes, Galileo rejects this and suggests rather that

"The present does not seem to be the proper time to investigate the cause of the acceleration of natural motion, concerning which various opinions have been expressed by various philosophers, some explaining it by attraction to the center, others to repulsion between the very small parts of the body, while still others attribute it to a certain stress in the surrounding medium which closes in behind the falling body and drives it from one of its positions to another. Now all these fantasies, and others too, ought to be examined; but it is not really worth while. At present it is the purpose of our Author merely to investigate and to demonstrate some of the properties of accelerated motion (whatever the cause of this acceleration may be) . . ."

The primary question has changed from "why" to "how," and to imple-
ment this Galileo turned to a mathematical description of natural phenomena.

In the course of his investigation Galileo wrote the equivalent of a modern scientific monograph. First he stated his intention — to set forth a new
science dealing with an ancient subject: change in motion. In the discussion of free fall proper, Galileo noted that it was well known that bodies accelerate as they fall. What was to be determined was just how this acceleration occurs. At this point he introduced definitions he planned to use
(including those of "uniform motion," "velocity," and "uniformly accelerated motion"). The reader is next informed that Galileo will limit his discussion to falling bodies alone: "We have decided to consider the phenomena of bodies falling with an acceleration such as actually occurs in nature."
Note how different this is from the Baconian method, in which all examples of motion would have been gathered prior to the determination of
scientific laws.

At this point Galileo introduced a rule of simplicity before proceeding
further: "Why should I not believe that such increases [in velocity] take
place in a manner which is exceedingly simple and rather obvious to every-
body?" That is, if bodies accelerate in free fall, we might make the assumption that they accelerate in the simplest possible way, uniformly. A test
now seemed imperative and in the dialogue Galileo's friend, Sagredo, admitted that

"I can offer no rational objection to this or indeed to any other definition
. . . [but] . . . I may nevertheless without offense be allowed to doubt
whether such a definition as the above corresponds to and describes that
kind of accelerated motion which we meet in nature in the case of freely
falling bodies."

Galileo's answer was to deduce a series of theorems required if free fall is actually a case of uniform acceleration. Included are the familiar equations
$s = \frac{1}{2} vt$ and $s \propto t^2$, where s is distance, v is velocity, and t is time. An experimental proof is offered through the inclined plane, which permits the
retardation of the descent motion so that both distance and time may be
measured. Using a water clock as a timing device, Galileo obtained results
that upheld his derived formula, $s \propto t^2$. Here then was a case of uniformly
accelerated motion, even though he readily admitted that it was not free
fall.

To proceed further, Galileo next made the assumption that a body that
falls through a height of an inclined plane attains the same velocity as one

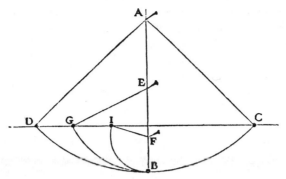

Figure 6.2. Galileo's pendulum experiment. From Galileo, Galilei, *Discorsi e dimostrazioni matematiche, intorno à due nuove scienze attenenti alla mecanica & i movimenti locali* (Leiden, 1638). Courtesy of The Joseph Regenstein Library, The University of Chicago.

that falls through its length. After presenting the logic for this, he again turned to experimental proof. In this case he turned to a pendulum, noting that it swung from a given height on one side to the same height on the other (after making some allowance for the resistance of air). But the motion of a pendulum describes the arc of a circle and therefore may be viewed as a descent along a series of inclined planes of different inclinations (Figure 6.2). These hypothetical planes were approximated by pegging nails in the board behind the pendulum. In all cases the pendulum bob came close to reaching the original height (therefore attaining nearly the original velocity on the descent). In this way it seemed possible to deduce that the times of descent along the lengths of inclined planes are in simple proportion to their heights, and that the accelerations are inversely proportional to the times of descent. That is, $\dfrac{t_1}{t_2} = \dfrac{h_1}{h_2}$

but $v = a$ (acceleration) t

or a $\propto \dfrac{1}{t}$

therefore $\dfrac{t_1}{t_2} = \dfrac{a_2}{a_1}$

All of this leads to the conclusion that free fall is uniformly accelerated. The result of course is basic for the physics of motion, but here it is of special interest to us for an example of procedural method. Galileo first stated his problem and then carefully enunciated the definitions he intended to use. Then he made a fundamental assumption regarding free fall and uniform acceleration that must be tested. This was done by checking theorems that must hold true if free fall is indeed uniformly accelerated. The experiment was then carried out and a further assumption made and tested before the original one was accepted.

All of this is carried on in the form of a dialogue, with lengthy digressions on the part of the participants. A weakness from our point of view is that the demonstrations are given in the form of thought experiments: that is, suggested experiments with no evidence of their actual completion. But the importance of it all is that here we see how a working scientist investigates a specific problem through the constant interplay of hypothesis and experiment. The suggested procedure is one that could be followed today. Indeed, many college students will recall that Galileo's inclined plane and pendulum are still used in laboratories as an introduction to the physics of motion.

Bacon and Descartes had both called for the destruction of the ancient philosophies. No one could deny their impact on the scientific world of the late seventeenth century, but in the end the influence of Galileo may well have been greater in the development of modern scientific method. The men of the Royal Society of London did not hesitate to designate themselves as "Baconian," but their Galilean heritage is no less evident.

Thought Experiments, Observations, and Atomic Theory

Galileo's study of free fall involved two fundamental experiments: the inclined plane and the pendulum. Yet, as presented to the reader, Galileo's work emphasized the logic of the conclusions rather than their experimental foundation. Accordingly, Galilean scholars have debated at length whether or not Galileo actually performed the experiments. Although this question need not detain us, Galileo clearly left himself open to criticism, a criticism that is evident in his discussion of the earth's motion. Here a question frequently asked was: Where would a stone fall when dropped from the mast of a swiftly moving ship? When at anchor it is obvious that it falls parallel to the mast, but when in motion one might conceivably expect it to fall far to the rear since the ship has moved ahead during the time of fall. The problem was of real interest to both Aristotelians and Copernicans, for both saw in the example a possible analogy with the moving earth. Galileo, employing the medieval term of "impetus," stated firmly that in the moving ship the stone continued to fall parallel to the mast, explaining that the stone partook of the forward impetus of the ship. In the dialogue Simplicio, the Aristotelian spokesman, replied that, "Not only haven't you made a hundred tests of this, you have not even made one." Quite so, replied Galileo's spokesman, "I am certain without experiment that the event will be as I have told you; for it must be so." The conclusion may have been correct, but the argument was not completely convincing in 1638.

The study of local motion was related not only to the fate of the Copernican system, but also to the revival of the atomic theory. In his own discussion of motion Aristotle had assumed that speed was inversely proportional to the resistance of the medium. Therefore in the case of a vacuum (no resistance) an infinite speed would be attained. This was unthinkable and therefore "nature abhorred a vacuum." Since a vacuum was impossible, it was also necessary to reject the atomic explanations that had been proposed by the pre-Socratic philosophers.

Aristotle had also suggested that bodies of different weights move in the same medium with velocities that stand to one another in the same ratio as their weights. For one who had observed objects falling in liquids and oils — or a stone and a feather dropped simultaneously in the air — this would indeed seem to be the commonsense answer. But what would happen in a vacuum, assuming that one could be produced? Galileo pointed out that the less dense a medium, the closer the velocities of the falling bodies become, regardless of their weight. Therefore, he argued, in a vacuum all bodies would fall at the same rate. Consequently the discussion of the relative velocities of falling objects of different weights was associated with the question of the existence of a vacuum, and far more important, the possibility of the atomic makeup of matter.

In the search for a mechanistic philosophy divorced from vitalistic and theological connections it is understandable that a naturalistic explanation of phenomena solely in terms of the size, shape, and motion of particles would be appealing. We have already noted Descartes's earlier particulate system, but for different philosophical reasons he had rejected the possibility of a vacuum. And although it might be unexpected, even the alchemists were using a blend of atomism and vitalism in their explanations by the early seventeenth century. But it was Galileo's belief in an atomic makeup of matter and the existence of a vacuum that was to prove most influential on this score. Accordingly, we find Evangelista Torricelli (1608–1647), a friend and disciple of Galileo, describing the mercury barometer shortly after his master's death. Later the study of the vacuum at the top of that tube was to occupy many hours on the part of the members of the Florentine Accademia del Cimento in the middle decades of the seventeenth century. In Germany Otto von Guericke (1602–1686) constructed his vacuum pump and demonstrated its effect in the Magdeburg hemispheres of 1657. And we find Robert Boyle commissioning Robert Hooke (1635–1703) to design an air pump in the 1650s so that he could carry out experiments *in vacuo,* which he proceeded to explain in terms of a corpuscular philosophy.

Galileo's discussion of a vacuum obviously had led to experimental work confirming his conclusions. But the same realization of the need for experimental proof will be seen if we return to the ship experiment. The argument of the moving ship had been one of the more powerful ones advanced by the opponents of Copernicus, and although Galileo had rejected their Aristotelian argument, he admittedly had not carried out the experiment to prove his case. Galileo's discussion had appeared in his *Dialogue on the Two Principal World Systems* in 1632. Prior to publication he had circulated his manuscript and it is clear from the ensuing correspondence that among those most interested were Mersenne and Gassendi. Mersenne immediately tried, in 1633, to determine the velocities of stones dropped from a cathedral tower. But, also seriously concerned about the ship experiment, he wrote to a friend who often crossed the English Channel. During a crossing made in 1634 this correspondent arranged to have a sailor climb the mast and drop weights. The results confirmed the work of Galileo, as the stones fell at the foot of the mast. But this work was not published, and it was Gassendi who next turned to the ship experiment. Convinced of its importance, he determined to have it carried out with an audience so that there could be no doubt as to the outcome. He found a patron in the new governor of Provence, who had an interest in the sciences.

An elaborate series of tests discussed by Galileo was planned by Gassendi and carried out in October, 1640. Men on horseback and in chariots threw stones in the air and across to each other, and it was seen that these missiles followed the forward motion of the horses as well as the transverse motion of the throw. Objects dropped from a horse in full gallop were found to fall in a straight line from the point of view of the rider. These and a series of other tests all confirmed the work of Galileo, but surely the most spectacular demonstrations were made aboard a naval trireme. After the ship had reached its highest speed it was found that, whether the stone was dropped from the mast or projected straight up, in both cases it fell to the foot of the mast rather than far behind in the stern. In his description of the results, Gassendi gave the speed of the boat and described all of the experiments in detail. He then suggested that the reader might carry out similar tests by throwing a ball in the air on his balcony while pacing – or by taking a small sailboat out on the river and testing the facts as he had on the trireme.

With Galileo we have a far different case than that of Kepler. The latter's work was barely known until a generation after his death, but Galileo's works were eagerly sought and read by a large contemporary audience. And when he admitted that he had not tried the ship experiment, there were those who realized that it had to be done — and that it had to be done with witnesses.

One senses a far different climate in the 1640s than that existing in the opening years of the century. With Bacon, Descartes, and Galileo criticism of Aristotle and the scholastic tradition had become far more sophisticated. And while we may point to flaws in their results as well as their methods, it is evident that their discussions regarding the need for a new science, the role of experiment, the proper use of mathematics, and the interplay of the inductive and the deductive processes in the role of discovery directly and dramatically affected the development of science.

Galileo's study of local motion rightly makes him a key figure in the rise of modern science not only for his methodology, but also for the consequences of his work. He had attacked a fundamental aspect of Aristotelian natural philosophy and had developed in its stead a new science of motion carefully built upon experiment and the consequences to be deduced from his evidence. Moreover, his rejection of Aristotelian local motion had reopened the possibility of an atomistic view of matter. This was a highly desirable result for those who wished to reject both vitalistic and mystical explanations. Indeed, corpuscular or atomic explanations were to become an integral part of the mechanistic philosophy. It is understandable then that it should have been a priest, Pierre Gassendi, who set himself the task of removing the taint of atheism that survived from the ancient atomistic texts.

The historian could easily content himself with tracing an account of the gradual decline of the authority of Aristotle in the course of the seventeenth century. However, such an account would not accurately depict the scholarly world of that period. Although Aristotle served as the whipping boy for many authors, a study of many texts frequently reveals a concern regarding influences other than scholasticism. Indeed, the newly rising mechanists were to find the chemists more dangerous than the declining proponents of the ancients, and it is to this problem that we turn next.

CHAPTER VII

The New Philosophy – A Chemical Debate

The style, brilliance, and influence of Bacon, Descartes, and Galileo may easily lull the modern historian into the belief that the vagaries of mystics and magicians were a thing of the past by the early seventeenth century. And it is true that we find widespread recognition of the need of a new science to replace that of the schools in the early decades of the new century. But to assume that the dreams of a "new philosophy" were to be limited solely to the mechanists would be far from correct. We need only return to the chemical philosophers or read the scientific utopias of the seventeenth century to see described a "new science" far removed from that of the mechanical philosophers. Bacon's *New Atlantis* is the best-known example of these utopias, but others by Tommaso Campanella (1568–1639) and Johann Valentin Andreae (1586–1654) reflect Hermetic themes in their ideal commonwealths. No less interesting is the Rosicrucian manifesto, which was an open statement calling for Paracelsian reforms in science, education, and medicine. And closely related are the works of Robert Fludd and Jean Baptiste van Helmont (1579–1644), which are of great interest in their assessment of the intellectual world of the period from the viewpoint of the chemist and the physician. The reaction to their work by contemporary mechanists attests to the continued authority enjoyed by hermeticists and chemical philosophers in the same period that saw the publication of the most important works of Galileo, Descartes, Bacon, and Kepler.

The Scientific Utopias

Bacon's *New Atlantis* was written late in his life (c. 1624) and published posthumously in 1627 by his friend William Rawley (c. 1588–1667), who explained that the work was meant to describe "a model or description of a college instituted for the interpretation of nature and the producing of great and marvellous works for the benefit of men." It is short — less than forty pages in the edition of 1664 — and it was widely read in the seventeenth century. There is little doubt that it was one of the chief sources of inspiration for the founders of the Royal Society of London.

The form of the New Atlantis bears a marked similarity to other works of this genre. A group of voyagers en route from Peru to Japan come across a previously unknown land, the inhabitants of which prove to be remarkably well informed both about nature and all aspects of the outside world. Offering medical aid to those voyagers who are ill, the inhabitants of this land, Bensalem, refuse any payment for their services, and in time they proceed to inform their visitors of their work and the secrets of their country.

Their primary concern is the search for a knowledge of heaven, but in practice this takes the form of the study of the world about them — a clear reflection of the contemporary search for God though his created nature. This research takes place in "Solomon's House" and we are told that "the End of our Foundation is the knowledge of Causes, and secret motions of things; and the enlarging of the bounds of Human Empire, to the effecting of all things possible." Here are to be found facilities for making all kinds of observations. There are deep caves where mining processes may be imitated and where experiments on the production of new metals may be carried out. Some study the curing of disease and the prolongation of life, whereas others are engaged in research related to the enrichment of the earth. There are towers a half mile high for experiments on refrigeration, as well as artifical lakes, wells, and foundations and parks containing all sorts of beasts and birds. A variety of furnaces, optical equipment, engines, and astronomical instruments permit all types of observations in chemistry, perspective, mechanical inventions, and astronomy.

The voyagers are told of the method employed by the scholars in Solomon's House. First the observations are collected and collated. Then various groups are assigned to draw conclusions and make suggestions for further study. When that complex process has been completed, there are "three that raise the former discoveries by experiments into greater observations, axioms, and aphorisms. These we call 'Interpreters of Nature.' " The entire process is, of course, the Baconian system in action. It is basically qualitative and observational, with little emphasis placed on mathematical interpretation. One would seek in vain for any directive to study the physics of motion and, indeed, the entire program places a heavy emphasis on what were then considered chemical studies and the beneficial form of natural magic.

A year prior to Bacon's composition of the *New Atlantis* there had appeared in Germany another scientific utopia, Tommaso Campanella's *City of the Sun*. This work was written in 1602 and it reflects Campanella's early devotion to Bernardino Telesio in his insistence that understanding proceeds primarily from the senses. But it also reflects the Hermeticism that formed an essential part of all his writings. *The City of the Sun* has political overtones as well. Hoping to establish an ideal commonwealth, Campanella had become involved in a plan to overthrow the Spanish domination of Naples in 1600. The resultant failure was to lead to more than twenty-seven years of imprisonment and torture. In his cell he wrote an astonishing number of books and manuscripts – and these were to include his description of the ideal City of the Sun.

The City of the Sun was a metropolis built on a hill, with seven concentric walls and a central temple. In many ways this plan is reminiscent of earlier portrayals of the temple of knowledge or the heavenly Jerusalem described in Book of Revelation. The walls were important to all citizens for thereon was displayed the wisdom of the world. Each was devoted to a different subject. The innermost depicted all mathematical figures on one side and gave a complete map of the world plus a description of all countries and their peoples on the other. The second wall was devoted to metals, stones, and minerals as well as natural and artificial liquids ranging from oceans to wines. Whenever possible, samples were embedded in the walls so that the populace might better understand the illustrations and diagrams. The next four walls displayed all types of animal and vegetable life, and the final one was devoted to the mechanical arts, with careful attention given to the great inventors. Here, among the lawgivers, were to be found religious figures: Moses, Osiris, Jupiter, Mercury, Muhammad – and Christ and his apostles.

The city's central temple had a vast dome mapping the heavens that was complete with verses describing the powers of individual stars. The city itself was ruled by a coterie of Hermetic priests, who led the populace both wisely and well through their knowledge of the stars and natural magic. As true natural magicians they knew how to employ the power of the stars for the benefit of those on earth.

If we were not already so familiar with this typically Renaissance mixture, Campanella's *City* might seem to be a strange confluence of magic and observation. In fact, there is no question about Campanella's interest in magic and astrology. In his *Metaphysica* (1638) Campanella readily admitted his espousal of the magic of Ficino, and this he ascribed ultimately to the texts of Hermes Trismegistus. There is also evidence that he engaged in magical practices.

But if Bacon may have known the work of Campanella, it is far more likely that he knew of the earlier works ascribed to the Rosicrucians. The two short texts that initiated what has been called the "Rosicrucian Furor" were the *Fama fraternitatis* and the *Confessio* (1614, 1615). Here was a Paracelsian call for a new learning couched in a utopian format. The reader was informed that the founder of their order, a certain fabulous Christian Rosenkreuz, had traveled as a pilgrim to the Near East where he had encountered learned communities in Damascus and Fez. Hoping to institute a similar group of European scholars dedicated to knowledge and service, he had returned home and gathered about him a small group of men who were inspired by his ideals. They had labored in secret throughout the life of their founder and long after, but at last it seemed to their successors that the time had arrived to announce their goals publicly.

In the *Fama fraternitatis* we find a call for a new learning to replace that of the universities. In place of Aristotle, Galen, and their later commentators scholars should seek out the truths of God and nature. In Paracelsian fashion, medicine was praised as the basis of all natural philosophy. For them it was a godly art those benefits should be dispensed without thought of payment. And although this true medicine was known to Christian Rosenkreuz, these Rosicrucians informed their audience that they knew of great physicians, philosophers, and magicians alive and at work in Europe. The greatest in recent times had been Paracelsus, and his works had been placed next to those of Christian Rosenkreuz in their secret vault.

This "Rosicrucian manifesto" displays a missionary spirit. The suggestion is made that great wonders might be accomplished if the truly learned scholars of Europe united for the benefit of mankind. However, if these scholars were not at the universities, where were they and how might they be contacted? The author suggested that they should declare themselves in writing and join the brotherhood in the forthcoming reformation of learning. For this reason the scholars of Europe must search their souls and "declare their minds, either *Communicatio consilio* or *singulatim* by Print." Both the *Fama* and the *Confessio* were to be published simultaneously in five languages so that no one could excuse himself and say that he had not seen the message — and while the brothers refused at that time to give out their names or announce their meetings, they were willing to assure those who answered their call that their works would not go unnoticed.

One might expect such short — and anonymous — texts to go unnoticed, but such was not the case. The *Fama* was actually published in four languages in nine editions between 1614 and 1617, and an English translation appeared in 1652. Letters still exist in European libraries from those offering to join the order, and one bibliographer has traced several hundred books and tracts that appeared within a decade debating the merits of this secret group. Major cities were visited by men who announced themselves as members of the brotherhood and promised to show all of their secrets to those who wished to become initiated. In an account published in 1619 we read:

"What a confusion among men followed the report of this thing, what a conflict among the learned, what an unrest and commotion of impostors and swindlers, it is needless to say . . . there were some who in this blind terror wished to have their old, and out of date, and falsified affairs entirely retained and defended with force. Some hastened to surrender the strength of their opinions; and after they had made accusation against the severest yoke of their servitude, hastened to reach out after freedom."

Since there is no indication that such a group ever did exist, the reaction is remarkable.

One notable product of this intellectual unrest was the utopian *Christianopolis* of Andreae (1619), who may be the actual author of the *Fama*. The *Christianopolis* bears striking similarities to the *New Atlantis* and its influence was almost as great inasmuch as it strongly influenced the English groups that formed the background to the Royal Society. In this work, too, we find the familiar account of the decay of European learning and religion coupled with the suggestion that a proper scholarly community be formed. Andreae's example is the ideal city of Christianopolis, where the citizens study both Holy Writ and nature. Books other than the Bible seem largely useless to the inhabitants because it is the study of nature that leads to a greater understanding of the Creator. "A close examination of the earth will bring about a proper appreciation of the heavens, and when the value of the heavens has been found, there will be a contempt of the earth."

Accordingly, the laboratory is of great importance to the citizens of the utopian metropolis. And, as we might expect, it is the chemistry laboratory, fitted out with the most complete equipment, in which "the properties of metals, minerals and vegetables, and even the life of animals are examined, purified, increased, and united, for the use of the human race and in the interests of health." More important, however, is the fact that here "the sky and the earth are married together," and the "divine mysteries impressed upon the land are discovered." These are clearly references to the macrocosm–microcosm analogy and the doctrine of signatures.

The importance of chemistry for Andreae becomes even more evident when one compares it with his treatment of other sciences. In the hall of physics the citizens of Christianopolis see painted scenes of the sky, planets, animals, and plants in a manner reminiscent of the concentric walls of Campanella. Here also may be examined samples of rare gems and minerals, poisons and their antidotes, as well as all sorts of things beneficial and injurious to the body. As for mathematics, the true student of Christianopolis may rise above vulgar arithmetic and geometry to contemplate the mystical numerical harmonies of the heavens known to the Pythagoreans of old. Everywhere the interrelation of heaven and earth is stressed and accordingly astrology is raised here to its proper place. For Andreae

"he who does not know the value of astrology in human affairs, or who foolishly denies it, I would wish that he would have to dig in the earth, cultivate and work the fields, for as long a time as possible, in unfavorable weather."

The implication is clear. A new learning is required, and if it could not be accommodated to the current university system, a separate academy or college must be founded. Andreae's proposals could have been seconded by any of the chemical philosophers.

Robert Fludd and Mystical Chemistry in a New Century

Among the mass of printed responses to the Rosicrucian manifesto are two pamphlets of considerable interest: one by Andreas Libavius and the other by Robert Fludd. We have already referred to Libavius (Chapter 2) as an anti-Paracelsian iatrochemist with a distaste for mystical interpretations of natural phenomena. He was convinced of the importance of using chemicals in medicine, but he wanted no part of a Paracelsian chemical philosophy that interpreted the universe in terms of the macrocosm—microcosm analogy. Accordingly he had supported the Parisian chemical physicians in their struggle against the Galenist-dominated faculty in 1606, and nine years later he condemned the Rosicrucian texts. To him these seemed to be both mystical and Paracelsian, and even worse, they promised the destruction of all aspects of ancient learning.

The attack of Libavius on the Rosicrucians was the cause of the first publications by Fludd, a knight and a man of substance who had studied at Oxford and had toured the major educational centers of the Continent. Elected a fellow of the Royal College of Physicians in 1609, he was later in contact with many of the most learned English scientists of his day. Fludd was a mystic at heart and, after learning of the *Fama* and *Confessio,* he wrote a reply to Libavius (1616) in which he attacked the study of the ancients in the universities and called for a new learning based upon religious truths. Arguing that there had been a decline in true knowledge since the time of Moses, Fludd suggested that in place of Aristotle and Galen the schools should turn to alchemy, natural magic, and a new medicine. He methodically criticized the liberal arts and specifically rejected the emphasis on logic in the scholastic curriculum. This, he felt, was reflected in the academic approach to mathematics, which was based upon definitions, principles, and discussions of theoretical operations. Rather, Fludd wrote, the mathematician should turn to the mystic teachings of the Pythagoreans, who had reached a certainty of belief in God through their study of numbers and their ratios. In this way he would be led to the concept of universal harmonies and to the very fabric of the world.

In his apology for the Rosicrucians Fludd insisted that we proceed to a new learning with a definite plan. He listed a series of key questions that should form the basis of future inquiries. We must, he wrote, consider the act of Creation by the divine light of the Lord. This, he affirmed, is no less than that vital spirit required for all life and motion. We must concern ourselves with all aspects of its action and pay attention to other concepts of interest — and here he included the atomic views of Democritus. And as we turn from the macrocosm to the microcosm, Fludd wrote that we must focus our attention on the assimilation of this vital spirit in the body. Here he emphasized that this spirit was to be found in the air and that it entered our bodies through inspiration. The need to determine just how the spirit nourished our bodies would make necessary a new study of the body itself. We must determine how the spirit is separated from the gross air, and how it is dispersed through the arterial and venous systems. One need go no further to see why Fludd was to take such an avid interest in Harvey's discovery of the circulation.

A second edition of Fludd's Rosicrucian apology and the first volume of his history of the macrocosm and the microcosm were published in 1617. The latter is probably the most detailed exposition of the two-world universe ever written, and for Fludd it seemed to fulfill his promise of a new science. The author considered his work, which is dominated by biblical and Hermetic citations, to be a true statement of the chemical philosophy. Beginning with macrocosmic events he discussed the Creation, the elements, and the order of the universe. Fludd militantly held to the geocentric universe, but he also described the "centrality" of the sun, arguing that it was situated midway between the earth and God. Other volumes dealt with the liberal arts, warfare, meteorology, anatomy, and medicine. The world described by Fludd emphasized universal harmonies, which he thought could be expressed in terms of Pythagorean number mysticism, as well as sympathetic action between the great and lesser worlds. Required by all was the vital spirit of the Lord, which descended from heaven and gave life to literally everything. Deeply convinced of the existence of this spirit, he described his attempt to extract it as a substance from wheat by chemical means.

All things were of interest to this man, who felt that his work would provide the needed basis for a Christian science that would replace the scholasticism of the universities. And, indeed, the scholarly community did read his work. We need not be surprised to learn that Kepler, Mersenne, and Gassendi were in the vanguard of those who saw in his work a threat to their own.

The Reaction to Fludd: Kepler, Mersenne, and Gassendi

The first major reply to Fludd's works was that of Johannes Kepler. In an appendix to the *Harmonices mundi* (1619) and in a later response to a reply by Fludd (1622) he examined the English physician's use of mathematics. As Kepler presented his case, the differences between his own approach and that of Fludd were simple, in our terms that of the "scientist" versus the "mystic." Kepler described his own concept of a universal harmony as "mathematical," whereas Fludd's explanations were "enigmatic, emblematic and Hermetic." How could any scientist compare Fludd's symbolism with his own true mathematical diagrams? And if Fludd reveled in his shadowy enigmas, Kepler had rescued the same phenomena from darkness and brought them into the light. Fludd, he added, had borrowed fables from the ancients whereas he had built upon the fundamentals of nature with mathematical certitude. Fludd had also confused things that he did not properly understand whereas he (Kepler) had proceeded in an orderly fashion corresponding to the laws of nature.

Perhaps we need not be reminded that in reality Kepler shared many mystical convictions with Fludd. But granted this, it is true that the meaning of mathematics for Kepler was quite different than it was for Fludd. The latter sought mysteries in symbols according to a preconceived belief in a cosmic plan. Consequently his proportions and harmonies were forced to fit these symbols. Kepler, perhaps just as obsessed with his own symbolic spherical picture of the world, insisted that his hypotheses be founded on quantitative, mathematically demonstrable premises. If a hypothesis, no matter how satisfying from a symbolic point of view, could not accommodate his observations, Kepler was willing to alter it. These two views were so opposed that the two men could not really understand one another. For Fludd Kepler was one of the worst sort of mathematicians, one of the vulgar crowd who "concern themselves with quantitative shadows." In contrast, he added "the alchemists and Hermetic philosophers . . . understand the true core of the natural bodies."

Although the Fludd–Kepler exchange is of considerable interest, the scope of the reaction to Fludd's publications among French scholars was to be much broader. The increased publication of alchemical and chemical texts, the announced "visitation" of the Rosicrucians to Paris (1623), and a widely publicized alchemical debate in Paris in the same year resulting in a series of arrests and the condemnation of the doctors of the Sorbonne all contributed to this new state of alarm.

Among the first of the French savants to reexamine the claims of the chemical philosophers was Marin Mersenne. In *La Vérité des sciences* (1625) he argued that a true science of nature would develop only after the mathematical interpretation of nature had overcome the claims of the chemists. In some detail he discussed these claims in dialogue form among an alchemist, a skeptic, and a "Christian Philosopher." For the alchemist no science was more certain than his own because alchemy taught through experience. To him there seemed little doubt that the works of Aristotle – admittedly filled with dangerous theological views – had been replaced by the sounder observational approach of the chemist.

Mersenne's rejection of the views of the alchemist was firm. For the "Christian Philosopher" the recent condemnation by the Sorbonne had been just. These learned doctors had rightly questioned the theological implications of the alchemical theses. Among them he noted the alchemical espousal of atomism, which at this early date Mersenne viewed as a position that might easily be overturned. As for the alchemists' much-vaunted "observationally" based system of elements and principles, Mersenne replied that the Paracelsian principles might yet be decomposed artificially. Should this occur, these principles need no longer be considered to be elementary.

And yet, Mersenne continued, if alchemy may be faulted on some points, it must not be rejected altogether. Rather, some method of control must be sought to avoid the dangerous pitfalls into which alchemists had fallen too often in the past. Mersenne suggested the establishment of alchemical academies in each kingdom that would take as their goal the improvement of the health of mankind. These academies would police the field not only by punishing charlatans, but also by actively engaging themselves in the reform of science. Allegorical and enigmatic terms must be discarded and replaced by a clear terminology based upon chemical operations performed in the laboratory.

For Mersenne a reformed alchemy would steer totally clear of religious, philosophical, and theological questions. It seemed to him that the subject served as a counterchurch for some, who argued that the most ancient theology, magic, and pagan fables were best explained through this science. Many, indeed, held to the chemical interpretation of the Creation. These dreams and speculations must be rejected at once if the subject was to gain the approval of the Catholic Church.

In his works Mersenne referred to a number of chemists whose publications he considered dangerous. One name that stands out is that of Robert Fludd, whom he branded as a heretic and magician of the worst sort. Deeply wounded, Fludd replied to the French monk in two works that restated his position on the chemical philosophy. Here he described once more the analogy of the macrocosm and the microcosm, the harmony of the two worlds, the significance of the vital spirit and its dispersal through the arterial system. True alchemy, Fludd insisted, has as its goal the establishment of the entire chemical philosophy as a basis of explanation for both man and the universe.

It is clear that Fludd's understanding of an *alchemia vera* was precisely what Mersenne found objectionable. Above all, Fludd was disturbed by Mersenne's warning that alchemists should disassociate themselves from religious matters. On the contrary, he felt that chemists and theologians had a common subject to investigate, namely, a part of practical theology that "we think to be nothing else but mystic and occult Chemistry." This subject is one that seeks to comprehend the Creation and the spirit of life. Nature and supernature are clearly united — and chemistry serves as a key to both.

Late in 1628 Mersenne sent a collection of Fludd's works to his friend Pierre Gassendi with an appeal for aid. In little more than two months the latter had completed his critique. Predictably, Gassendi rejected Fludd's explanation of the elementary principles and the Chemical Creation. And when confronted with Fludd's rejection of Copernicus and Gilbert, he could only conclude that "he understands another non-volatile earth and central sun than that commonly understood by us." Discussing the distinction made by Fludd between true and false alchemy, Gassendi complained of an interpretation that would make "alchemy the sole Religion, the Alchemist the sole Religious person, and the tyrocinium of Alchemy the sole Catechism of the Faith."

We need not follow this debate in more detail other than to recall that in the course of his reply Gassendi described and rejected Harvey's views on the circulation and that it was this that had led Fludd to defend his friend. Fludd's reply to Gassendi in 1633 was to bring a renewed effort on Mersenne's part to discredit him and the chemical philosophy. His correspondence — until well after the death of Fludd in 1637 — shows an unending effort to enlist the scholars of Europe against Fludd's dream of a "new philosophy" that was so opposed to his own.

The New Philosophy of Jean Baptiste van Helmont

In his search for support against Robert Fludd, Mersenne had written to many European scholars. Among them was Jean Baptiste van Helmont, who was to carry on a correspondence with the French savant. In one of the earliest letters van Helmont answered a query on the value of Gassendi's then-recent reply to Fludd. The Belgian physician-chemist replied flatly that Fludd was a poor physician and a worse alchemist – a superficially learned man on whom Gassendi should not waste his time. This appraisal is of great interest in that much of van Helmont's work was characterized by concepts and attitudes that had been condemned by Mersenne. Nevertheless, van Helmont's work seemed sufficiently different at the time for it to become the basis of a new iatrochemical school in the seventeenth century.

Van Helmont's search for truth was an intensely personal one. It had led him to reject a master's degree from Louvain because he felt that he had learned nothing there, and he was later to turn down the offers of employment from princes, preferring to devote himself to research at his home. Having little interest in personal fame, van Helmont put nothing in print until lured into a controversy over the weapon salve by a Jesuit in 1621. The belief that treatment of the weapon causing an injury would heal the wounded person was widespread in the seventeenth century and was based upon the concept of universal sympathetic action in nature.

Van Helmont's tract on this subject attacked all those who had engaged in the debate. He did not deny the efficacy of the cure, but he did take exception to those who had described it in supernatural terms. Insisting that this was a purely natural phenomenon, van Helmont went on to state that "Nature . . . called not Divines for to be her Interpreters: but desired Physitians only for her Sons." Indeed, one might think that it was Galileo rather than van Helmont who admonished his Jesuit adversary to "let the Divine enquire concerning God, but the Naturalist concerning Nature."

The weapon salve could be explained through the proper understanding of the harmony of the greater and lesser worlds in that "all particular things contain in them a delineation of the whole universe." As for Paracelsus, his works were to be praised and his three principles were to be accepted without reservation. Magic is "the most profound inbred knowledge of things" and its basis remains the same whether used for good or evil. Indeed, when one properly understands sympathetic action in nature one becomes aware that the effect of sacred relics differs little from the magnetical weapon salve itself. On this point van Helmont expressed a potentially dangerous stand for a professed Roman Catholic.

The publication of this tract could hardly have appeared at a less opportune moment. Van Helmont's attack on a prominent Jesuit and his defense of magic and Paracelsus, combined with his interpretation of the miraculous power of relics, could not go unnoticed. He was denounced by the faculty of medicine at Louvain in 1623 and shortly thereafter summoned to appear before the Spanish Inquisition. There numerous propositions from his work were proclaimed heretical and he was confined to prison, and later, house arrest. He was forbidden to publish anything without the consent of the Church. Even after his release in 1636 Church proceedings continued nearly until his death in 1644.

Van Helmont bequeathed a mass of manuscripts to his son for eventual publication. The complete works, the *Ortus medicinae,* appeared four years after his death and by 1707 had been printed twelve times in five languages. In this highly influential publication we again meet with a strong plea for reform. It was necessary to "destroy the whole natural Phylosophy of the Antients, and to make new the Doctrines of the Schooles of natural phylosophy." Van Helmont characterized ancient science and medicine as "mathematical" and logical and he argued that this must be replaced by a truly observational approach to nature. Nor was the ancient approach to motion any better. Aristotelian local motion had led to a belief in an immovable mover. A Christian definition, van Helmont countered, would permit no such restriction on the Creator. In reality motion was inherent in life, implanted in the initial seed by the Creator. If mathematical abstraction could lead to such a mistaken conclusion, it could readily be seen that the Aristotelian description interpretation of nature

"is a Paganish Doctrine drawn from Science Mathematical, which necessitates the first Mover to a perpetual unmoveablenesse of himself, that without ceasing he may move all things . . . Therefore let the Schooles know, that the Rules of the Mathematicks, or Learning by Demonstration do ill square to Nature. For man doth not measure Nature; but she him."

Clearly the new philosophy envisioned by van Helmont was one that would reject any concept of nature interpreted primarily through mathematics.

Throughout van Helmont's work may be noted the close association of nature and religion. Once again we are told to look first at the account of Creation in Genesis. This, as in the case of Fludd, presents the order of Creation and the true elements. Here fire is not mentioned, whereas earth is seen as the product of water. As for the Paracelsian principles, they are useful since they are obtained by distillation from so many substances, but the more mature van Helmont no longer considered them to be truly elementary in nature. The key to nature is to be found in fresh observations, and it is chemistry that offers us our greatest opportunity for truth (Figure 7.1). Coupled with this, an awareness of quantification – understood here in terms of laboratory weights and measurements rather than mathematical abstraction – might well offer new insight. Van Helmont sought to demonstrate the elementary nature of water by weighing a willow tree in various stages of its development, but he was also interested in the determination of the specific gravity of metals and the comparison of the weights of equal volumes of urine as a guide to illness. He sought more accuracy in the determination of a scale of temperatures and was led by his studies to insist on the indestructibility of matter and the permanence of weight in chemical change.

A thoroughgoing vitalist, van Helmont proceeded to develop an explanation of the existence of all things based upon his system of elements and their own life cycles. Here he discussed life sources and seeds, which were responsible for results as diverse as the minerals and human disease. His medicine reflects his overall philosophy. Unwilling to accept the ancient medical texts, he was also disturbed by those who would accept everything ascribed to Paracelsus. Thus, in these later works van Helmont rejected the doctrine of the microcosm that postulated man as an exact replica in miniature of the greater world. Still, this did not prevent him from calling attention to numerous similarities between man and nature as a whole. Nor was van Helmont less concerned than Fludd with the vital spirit in nature. Fludd had sought to isolate this spirit through an alchemical experiment on wheat; van Helmont sought to do the same through the distillation of blood. His deep belief in the existence of this vital force in the blood caused him to be one of the first physicians to argue against bloodletting.

Many other examples might be brought forth to indicate van Helmont's interests, but it is more important here to emphasize that although both Fludd and van Helmont were influenced by their Hermetic and Paracelsian background, there were deep differences between these two chemical philosophers. The first, inspired by the utopian Rosicrucian manifesto, had sought a new approach to nature stressing inner truths, true religion, and a

Figure 7.1. The chemical philosopher as imitator and interpreter of natural phenomena. From J. B. van Helmont, *Opera Omnia* (1682). From the collection of the author.

mystical alchemy. The second may not have materially disagreed on these points, but he was to place a much greater emphasis on new observations.

Regardless of the fact that they shared many beliefs, Fludd and van Helmont were viewed as very different by most seventeenth-century scholars. For many in the middle decades of the century van Helmont seemed to present a plan for a new philosophy fully as promising as that of the mechanical philosophers. Here was a "Christian" observational approach to nature that seemed to reject the mysticism of earlier Paracelsians but still indicated the validity of comparisons made between men and nature. In England van Helmont's work inspired Walter Charleton (1619–1707) to prepare a number of translations of individual tracts by van Helmont in 1650, and Robert Boyle repeated the Helmontian arguments against the Paracelsian principles in *The Sceptical Chymist*. Thomas Willis (1621–1675) drew on the *Ortus medicinae* as the basis for his new chemical physiology, and even Isaac Newton read the work with care, taking extensive notes on its contents.

The debates centered on the work of Fludd and the broad contemporary interest in the work of van Helmont indicate the wide concern aroused by the chemical philosophy in the seventeenth century. Fludd's confrontations with Kepler, Mersenne, and Gassendi began with his earliest publications (1616, 1617) and continued for twenty years. Van Helmont's problems also began with his first publication (1621), but due to official persecution relatively little was known of his views until the posthumous publication of his collected works in 1648. Thus the European scholarly community was faced with a new, more observationally oriented chemical philosophy just when it was beginning to assimilate the work of Descartes and Bacon. Van Helmont's pleas for educational reform, his rejection of ancient philosophy, and his many observations were noted by a wide spectrum of European scholars at that time.

Reference to the chemical philosophers occurs frequently where it might at first be unexpected. However, this is almost inevitable, inasmuch as the chemical philosophers did not think of their work simply in terms of chemistry or medicine. Rather, this was a professed attempt to found a *philosophia nova* that would account for the entire cosmos. Only when approached in this fashion is it possible to explain the threat to natural philosophy seen in Fludd's mathematics by Mersenne and Kepler or the widespread interest and influence of the work of van Helmont. Indeed, only when understood in its seventeenth-century context — as a blueprint for the "new science" — may we expect to find anything here that could conceivably be of interest to the Isaac Newton with whom we are most familiar.

CHAPTER VIII

Epilogue and Indecision

To some it may seem misleading to close a work on Renaissance science and medicine with a discussion of Robert Fludd and Jean Baptiste van Helmont. With these authors we find far less scientific "progress" than we might wish. But to have turned last to Fludd and van Helmont does emphasize the complex nature of the period. Historians in the past have rightly concentrated on the rise of the mechanical philosophy in the seventeenth century, but the failure to have simultaneously assessed the reasons for the contemporary appeal of Paracelsism and natural magic has frequently resulted in an incomplete appraisal of the period as a whole. And again, it is only recently that full weight has been given to the fact that some traditional heroes of the new philosophy adhered staunchly not only to aspects of magic and mysticism, but also to key tenets of ancient philosophy. If the scholars of the seventeenth century had not considered the work of Fludd and van Helmont important, then it would never have become a center of debate. The same argument holds true for the case of Paracelsus a century earlier.

Science and the Two Humanisms

It is not too difficult to see why positivistic histories predominated in the past. Science and medicine were transformed in the period of the Renaissance. The world of the mid-fifteenth-century scholar had been largely interpreted in terms of medieval scholasticism. To be sure, freedom to criticize had affected key areas of thought. Thus a close examination of local motion by fourteenth-century scholars made evident the weaknesses present in Aristotle's position, and the importance of observational instruction in anatomy was rapidly recognized and led to the widespread accep-

131

tance of public dissections for medical students. These developments may have occurred only in a limited number of universities, but these were among the most prestigious centers of learning in Europe.

The internal self-criticism of the late medieval period was to be replaced by open revolt in the course of the next two centuries. A series of fundamental breakthroughs in the physical and biological sciences had occurred, and by the mid-seventeenth century most scientific and medical research was to be found outside the university. If it would be incorrect to suggest that the universities played no part in these developments (Padua and the medical tradition being the most notable exceptions), it is surely true that by this time the early scientific academies and local groups of scholars were playing a more significant role than the traditional educational centers. The now-strident call for a new science or philosophy was heard, demanding a replacement for the Aristotelian and Galenic training, which seemed useless, mentally stifling, and (at times) theologically suspect.

Any study of Renaissance intellectual history – and both the history of science and medicine should be covered by this blanket term – must take into account the influence of humanism. We have noted that literary humanism was late in reaching the sciences. Only in the second half of the fifteenth century do we find in scientific circles the same devoted search for texts of the ancient classics that had been part of the literary scene for well over a century. Thus it was in the final decades of the fifteenth century that we found Peuerbach and Regiomontanus seeking a complete Greek text of Ptolemy's *Almagest*. In medicine the similar endeavors of Linacre and Guinter of Andernach came even later. The work of these scientific and medical humanists played an essential role in the development of modern science. Copernicus was truly a product of the Ptolemaic revival, and no less did Vesalius and Harvey reflect sixteenth-century Galenism.

This form of humanism was an influential part of the background to the *De revolutionibus*, the *De fabrica*, and the *De motu cordis*. The study of the *Almagest* by Copernicus had transformed that work into the basis of a new world system, but in his hands the structure of ancient astronomy had remained clearly visible. The problems associated with Copernicus's concept of a moving earth were to lead to a new physics of motion and the difficult questions raised by the possibility of an infinite universe. No less significant was medical humanism, which may be followed through the Paduan tradition down to Harvey's discovery of the circulation of the blood. In the same fashion as the literary humanists these scholar-scientists and physicians revered the ancient authorities. But their very work was to lead to the destruction of ancient authority. The requirements of an earth in motion necessitated a physical system far different from that of Aristotle, and Galenism could no longer dominate a medicine transfigured by the new physiological discoveries of the seventeenth century.

The same discoveries were eventually seen to have theological overtones, which were seldom remarked by their authors. If Copernicus had been attacked by Lutheran divines for his heliostatic world system, he had been honored by his own church. By the early seventeenth century the situation had altered radically. Galileo was first warned and then prosecuted by the Inquisition. And Descartes, fearing the consequences of his devotion to the Copernican cause, hastily recalled his *Le monde* from the printer in 1633. If by this time religious fundamentalism required a literal interpretation of Holy Writ, scientists replied that the Scriptures were not meant as a guide to the study of nature. Deism was still far in the future, but its seeds were already sown.

Renaissance humanism was far more complex than might appear at first glance. Not only was there a recovery of the classics of ancient medicine and physics, there was also a recovery of the mystical texts of late antiquity and these were to be no less influential than Galen and Ptolemy. Marsilio Ficino translated and studied the mysteries of the Hermetic corpus and Plato's *Timaeus* in an effort to uncover the hidden relationships of the macrocosm and the microcosm. The result was a surge of new interest in natural magic and all its allied fields. Students of astrology, alchemy, the cabala, and Pythagorean numerology vied with each other in their search for a new key that would unlock the mysteries of the universe.

This humanistic strain, Hermetic, magical, and alchemical, was deeply ingrained in the science of the period. These scholar-mystics continually repeated their belief that man must study God's Creation so that he might better understand the Creator himself. True science and medicine for them was nothing but the knowledge of the secrets — and the hidden powers — of nature. In short, science and medicine were both seen as aspects of natural magic. Man would learn by his observation of those essential harmonies that linked all parts of nature. Agrippa, Porta, and Dee all participated in this mystical search for truth through nature. But the most influential of all were the Paracelsians, who openly called for the destruction of ancient authority. They were the ones — rather than the Paduan anatomists or the Copernicans — who saw the immediate need for a new and a different science and medicine. And their conviction that this would be based on their own medico–chemical system resulted in a debate that was as acrid as it was significant.

One must thus begin by acknowledging that the scholars of that period were well acquainted not only with Euclid, Aristotle, Hippocrates, Ptolemy, and Galen, but also with the Hermetic corpus and the works of alchemists and astrologers. There surely was a widespread call for a new philosophy, but again, this was the dream of Paracelsus, Campanella, and the Rosicrucians no less than it was that of Bacon, Descartes, and Galileo. And if we point to the emergence of mathematical abstraction and quantification essential for the development of modern science, this did not seem as significant then as it does now. For many at that time, a return to "true" mysticism and to natural magic seemed far more important. Nowhere is this "other road" to a new philosophy more in evidence than in the scientific utopias of the early seventeenth century.

It is the constant interplay of the "rational" and the "irrational" that continually redirects our attention to the debates of Robert Fludd and to van Helmont's pleas for reform. Fludd was the spiritual descendent of the sixteenth-century Renaissance Hermeticists, whose work was to become a storm center of debate. The chief attacks on his work were made by Kepler, Mersenne, and Gassendi, scholars who were primarily mathematicians, astronomers, and physicists. We may then view the exchange between Fludd and his opponents as a late expression of an ongoing debate resulting from the traditions of Renaissance literary humanism and Renaissance Hermetic humanism.

As for van Helmont, he was no less committed to the replacement of ancient philosophy with a new science than many of his contemporaries, but he openly opposed the more occult works of Paracelsus, the collected writings of Fludd, and all those alchemical and iatrochemical authors who emphasized mysticism and magic. Thus, in spite of their general agreement on the need for a chemical philosophy, Fludd and van Helmont differed noticeably in their approach. It is of interest to see that the first was bitterly attacked by the early mechanists, whereas the second was read with care and appreciation. A deeply religious man, van Helmont had nevertheless demanded a new philosophy divorced from the control of the Church. Indeed, van Helmont was to become recognized as the founder of a new approach to the study of nature that was to rival that of the mechanical philosophers in the third quarter of the seventeenth century. Prosecuted by the Inquisition and placed under house arrest, his case parallels that of Galileo in a number of important ways.

The Betterment of Man: Education, Agriculture, and Warfare

By the mid-seventeenth century there were many who spoke openly of the debate between the ancients and the moderns, but who understood fully that the problem was very complex. As we have suggested, the most critical debate may well have been between the mechanists and the chemists. This is best illustrated in the English literature from the middle decades of the seventeenth century. The Civil War and the Interregnum had encouraged a broad spectrum of proposals for reform within the Commonwealth. Among these were a number of plans dealing specifically with educational reform. Here were to be found demands for a complete change in the university curricula. Helmontians insisted on a higher education based upon new observations from nature to accompany the religious reforms resulting from the political revolution.

Perhaps the most interesting of these proposals was that made by John Webster in 1654. Originally a Puritan and later a Nonconformist minister, Webster studied alchemy as a young man and knew enough medicine to serve as a surgeon — as well as a chaplain — in the armies of Oliver Cromwell. Convinced that needed religious reforms were being accomplished in the early years of the Commonwealth, he became increasingly more distressed with the training of ministers at the universities. Accordingly, he wrote an *Examination* in which he demanded educational reform in terms that recall the words of both Fludd and van Helmont.

Reacting against the sterile and, to his mind, atheistic writings of Aristotle taught at the universities, Webster favored instead the "highly illuminated fraternity of the Rosie Crosse." As the Paracelsians asserted, true Christian knowledge of nature would be taught best by ocular demonstrations learned by putting "hands to the coals and furnace." In this way we might learn the importance of the three principles while we continued to seek out the secrets of natural magic and "Cabalistick Science." In general if we were properly to reform our knowledge on Christian principles we must seek to build up tables of axioms as Bacon had suggested, but we must also — as true Christians — seek a knowledge of nature

"that is grounded upon sensible, rational, experimental, and Scripture principles: and such a compleat piece in the most particulars of all human learning . . . is the elaborate writings of that profoundly learned man Dr. *Fludd,* than which for all the particulars before mentioned . . . the world never had a more rare, experimental and perfect piece."

In addition, the practitioners of the new philosophy were told by Webster to avoid Aristotle and to turn to the works of Ficino, Plato, Gilbert, and Hermes Trismegistus, as interpreted by the Paracelsians. Experimental chemistry was to be the key to nature, and the medicine of Paracelsus and van Helmont was to replace that of Galen.

Webster's tract is of special interest because it was scathingly attacked by Seth Ward (1617–1689) and John Wilkins (1614–1672), who are both revered today as founding fathers of modern science. The first had taught astronomy at Oxford since 1649, where he had introduced the teaching of the Keplerian elliptical orbits in England. Wilkins was also well known as the author of a defense of the Copernican system, as well as the *Mathematical Magick* (1648) and a semiutopian *Discovery of a World in the Moone* (1638). Ward and Wilkins were now to collaborate on a *Vindication* of the universities. Here Ward emphasized the high level of scientific work at the universities, and the impropriety and inconsistency of Webster's suggestions. In a point-by-point rebuttal of Webster's work, Ward admitted that the state of mathematics at the universities could be bettered, but hardly through the examination of mystical harmonies as proposed by Webster. He questioned the wisdom of completely discarding the Galenic medicine, and he argued that Webster was unfair in his charge that chemistry was unknown at the universities. And if Webster had pointed to the writings of Francis Bacon as basic for a reform of nature, Seth Ward agreed, but how did Webster follow this suggestion? He had turned to the "rational" and "experimental" Robert Fludd. Surely this was too much! "How little trust there is in villainous man!" Although a moment before he had recommended Francis Bacon "for the way of strict and accurate induction," now he

"is fallen into the mystical way of the *Cabala,* and numbers formall: there are not two waies in the whole World more opposite, then those of the L. *Verulam* and D. *Fludd,* the one founded upon experiment, the other upon mysticall Ideal reasons; even now he was for him, now he is for this, and all this in the twinkling of an eye, O the celerity of the change and motion of the Wind."

And Ward concluded, if Webster had then discussed Plato, Democritus, Epicurus, Philolaus, and Gilbert, why should there have been any need in this, for "if *De Fluctibus* be so perfect, what need we go any farther?"

The Webster–Ward debate is often misunderstood. John Webster has been accused of rashly proposing an odd mixture of science and superstition – chemistry and magic – as a basis for university reform. Ward is praised for reacting with righteous anger. But as we have seen, Webster's view of chemistry is not what we mean by modern scientific chemistry, and his magic is far from the black art the word may imply to us. Here he would have agreed with Bacon that the term "natural magic" had long been misapplied, and that it really signified "Natural wisdom, or Natural Prudence . . . purged from vanity and superstition." Adding to the complexity is the fact that when Thomas Hall (1610–1665) decided in the same year to defend the Aristotelian position in another answer to Webster, he stated that it was unnecessary to discuss the state of the sciences as Ward and Wilkins had admirably presented this case. Thus we find the champion of the ancients allied with the mechanists – or "Moderns" – against the chemical philosopher.

The search for educational reform was common to chemists and mechanists alike, but by the mid-seventeenth century the two groups saw their positions as fundamentally opposed. Of the two it was certainly the chemists who were most interested in a radical change. But also common to both was the desire to use the new philosophy for the benefit of man and the Commonwealth. This surely derives from the goals of natural magic – as opposed to black magic. The natural magician sought the hidden secrets of nature in order to apply them for a practical goal. On the microcosmic level this had encouraged the Paracelsians in their pursuit of a new medicine, although it was to lead to results in other fields. For Francis Bacon natural magic was nothing if not practical in its nature. He insisted on a survey of the arts and crafts so that these might be improved through further study. His hopes for separate "histories" of each were to become a characteristic part of the program of the early Royal Society of London and all other seventeenth-century scientific academies as well.

The early issues of the *Philosophical Transactions of the Royal Society of London* (beginning in 1665) reflect the practical goals of Bacon. Alongside purely scientific communications are medical reports for the benefit of physicians. As early as 1665 a special committee was established to examine the state of agriculture and to suggest possible improvements. These men placed special emphasis on normal farm products, but the society also encouraged its members and correspondents to communicate their observations on more exotic plants that were grown in other parts of the world. They also encouraged those concerned with navigation. Thus we find detailed descriptions of new timekeeping devices (for the determination of longitude), as well as observations on magnetic variation in different parts of the globe. Mining was considered vital; there are many reports on mines and ores from all parts of Europe. Similarly all kinds of new products and manufacturing processes were discussed. An improved plow, a speaking trumpet, building stones, the manufacture of iron, the casting of metals, the brewing of beer, and many other subjects of this sort were described in detail. There is little doubt that Bacon would have approved.

This practical bent has long been noted as a chief characteristic of the new science. But were the chemical philosophers any different from the

Baconians in this goal? They were not. Their medicine was based on practice and they openly scorned the theoretical physicians of the schools. Van Helmont challenged the Galenists to a test that would decide the future course of medicine:

"Let us take out of the hospitals, out of the camps, or from elsewhere, 200, or 500 poor people, that have fevers, pleurisies, etc. Let us divide them into halfes, let us cast lots, that one halfe of them may fall to my share, and the other to yours: I will cure them without bloodletting and sensible evacuation . . . [and] we shall see how many funerals both of us shall have."

His followers repeated the challenge for decades but no takers were to be found.

For others chemistry was seen as the potential savior of the state through agricultural reform and a new chemical warfare. Paracelsus had indeed written of a life-giving salt in manure as the reason for its use in fertilization. This theme was quickly seized upon by the French Paracelsist Bernard Palissy (c. 1510–1589) in his discussion of the use of marl as a fertilizer. By the end of the century Hugh Plat (1552–1608) in England had published a lengthy account of agricultural practices using the newer fertilizers. For him the benefits that would accrue to the nation were incalculable. His account repeatedly employed the Paracelsian life spirit and chemical theory in a discussion of the problems of generation and growth. Similar experiments are next to be found in the work of Francis Bacon — and we may thus not be too surprised to find that one of the earliest of the Baconian "histories" attempted by the Royal Society was that of agriculture. And if the full results of this committee were never published, the reports that do remain indicate once again that the subject was pursued through the traditional chemical approach.

The chemists' interest in chemical warfare is seen best in the work of Johann Rudolf Glauber, whose authority among chemists in the second half of the seventeenth century was considered second only to that of van Helmont. Deeply and personally affected by the tragic events of the Thirty Years' War (1618–1648), Glauber wrote of the need for the preservation of law and order but no less important to him was the establishment of Germany as the "Monarch of the World." This end was to be accomplished partially through a new economic prosperity and partially through a new military technology. Both were to be the result of the proper application of chemical knowledge gained from the chemical philosophy.

Economic prosperity was to be achieved through a new attention to the chemical philosophy. Farmers who had been ruined through years of overabundance and years of want might learn to concentrate their excess grain in good years to a liquid malt, which might later be reconstituted to beer. Grape farmers might benefit from the same process. In this case the must was to be evaporated to the consistency of honey, which could be preserved as long as needed and then reconstituted with water prior to the fermentation state. Not only would the vintner save carriage costs if shipment was necessary, but he might also withhold his product from the market until a time of need and thus reap an appreciably greater profit.

Farmers might then assure their own prosperity by wasting nothing. But the late war had shown that wise agricultural management would count for little against the destruction perpetrated by soldiers on the move. The state must be able to protect its people, their land, and their possessions. And if there now was peace one could not tell how long this might last. Even as Glauber wrote, he heard that the Turks were on the move — and recent unnatural storms and earthquakes and an unusual comet (1662) surely indicated divine displeasure. To Glauber this all meant another disastrous war was imminent. The state must therefore arm itself with new weapons that he himself had developed. He described long "warlike canes" through which acid could be projected in the form of a mist or rain. In this way the defenders of a besieged city might blind their adversaries. Or, if Christian forces were on the offensive, small hand grenades filled with the same acid liquid might be used to blind the defenders of the watchtowers of the enemy stronghold. The city gates might then be opened so that the Christian army might enter.

Glauber was fully aware that there was a moral question involved. Some had been repulsed by his work because through it others were to be maimed. Glauber answered that there was a great difference between gunpowder and his acids. With the former "a multitude of Men are destroyed and slain."

"But by this Invention of mine, no man is slain, and yet the victory wrested out of the Enemies hands. And the Enemies being taken alive and made Captives, may be constrained to work, and in my opinion may bring more benefit than if they were slain."

In any case, "is it not lawfull for us to smite our Capital Enemies the *Turks* with blindness, and to defend our selves, our Wives and Children?"

Fully as interesting is the fact that Glauber foresaw the possibility that some of these new weapons might be sold by traitors or fall into the hands of the enemy in the course of battle. It seemed almost inevitable to him that the secret would be lost in time. For this reason it was essential that

"Men of a quick piercing Wit" should constantly seek to improve existing weapons as well as invent new ones. Should such a program of research be instituted, "I do not question but that hereafter Wars will be waged after another manner than hath hitherto been done, and force must give place to Art. For Art doth sometimes overcome strength."

The examples of educational and agricultural reform and chemical warfare are fascinating examples of the practical goals that the chemical philosophers sought to accomplish. But the same examples have a still greater importance in that they indicate that even though these same chemists may have fought verbal duels with the mechanists and the early members of the scientific academies (to which they seldom belonged), they were no less concerned about the application of their knowledge for the benefit of mankind and their states than were their scientific and medical adversaries.

There surely was a Scientific Revolution. But as a revolution it was a long-term affair. The monumental changes we have chronicled took place over a period of centuries rather than decades. And for most of the period we have discussed, there was a constant dialogue and interplay between the intellectual descendants of Ficino and Paracelsus on the one hand and of Guinter of Andernach and Peuerbach on the other. The Hermeticist and the alchemist continued to debate his Galenist and Ptolemaic (or Copernican) adversary until well into the seventeenth century. It may be this continued exchange that best sets the limits of "Renaissance" science. Other aspects of the sciences could and did change, but they did not greatly affect this debate. Thus the growth of new scientific academies in the decades after 1660 may be seen as a realization of the dreams of the earlier scientific utopians. But if the Baconians and the mechanists were to dominate the membership of these organizations, the Royal Society did not exclude a prominent alchemist such as Elias Ashmole; as for John Webster, he was to write of the wonderful work of the Royal Society, which seemed to him to be fulfilling the dreams of the chemical philosophers.

Chronologically, we have discussed the period from the mid-fifteenth through the mid-seventeenth centuries — that is, from the translation of the Hermetic corpus and the work of Peuerbach down to the work of van Helmont and the early mechanists. But it would be incorrect to assume that all evidence of Renaissance Hermeticism disappeared in the course of the third quarter of the seventeenth century. The work of Boyle was strongly tinged with the earlier writings of van Helmont and he is only one of his generation who might be named. The best example may well be Isaac Newton, who was so deeply involved in alchemical studies that some scholars now insist that alchemy was the true basis of his physical theories. Such an assertion remains unproven; Newton's *Principia* (1687) is devoid of alchemical imagery and speculation. Newton's work represents the culmination of two centuries of debate over the true system of the universe and even today remains the foundation of the modern physical sciences. But for us Newton is of interest for a second reason: He published his experimental work on optics and his mathematical treatment of physical laws, but he filed away his alchemical manuscripts.

Newton's action is symbolic of the later history of science. The eighteenth century was the Enlightenment, the Age of Reason. Its science was "Newtonian" in that it was experimental science characterized by quantification and the use of mathematical abstraction in the description and clarification of natural phenomena. This was the science of the academies and the societies and it was a science that rejected and vilified the mysticism and magic so common to the Renaissance. But in fact alchemical texts continued to appear in the eighteenth century at a pace that rivaled that of the late sixteenth and the early seventeenth centuries. Although this is so, the earlier debate was no longer a vital one because these subjects no longer formed part of the scientific mainstream. Indeed, this exchange was now dormant and would be renewed – in a different form – only with the rise of *Naturphilosophie* at the dawn of a new century.

Suggestions for Further Reading

Literature on the Renaissance relating only to science and medicine has become both voluminous and highly specialized. For this reason the present essay is limited primarily to book-length studies in English. There are a number of exceptions, however, and the reader will find many articles and even a few larger works in other languages listed. If the emphasis has been placed upon recent studies in the field, real effort has nonetheless been made to indicate the wealth of primary source material that exists in English.

It would be impossible to prepare any such essay without paying tribute to Jacob Burckhardt's *Die Cultur der Renaissance in Italien* (Basel, 1860). There are many translations available in English as *The Civilization of the Renaissance in Italy*. Those interested in the philosophical background to Renaissance science should also refer to Paul Oskar Kristeller's *Eight Philosophers of the Italian Renaissance* (Stanford, 1964), *Renaissance Thought I: The Classic, Scholastic, and Humanist Strains* (New York, 1961), and *Renaissance Thought II: Papers on Humanism and the Arts* (New York, 1965). A short account of *Renaissance Humanism 1300–1550* has been prepared by Frederick B. Artz (Oberlin, Ohio, 1966), whereas an old, but sweeping, survey of *Thought and Expression in the Sixteenth Century* by Henry Osborn Taylor (2 vols., 1920; 2d rev. ed., New York, 1959) attempts to cover all aspects of intellectual life in that period. Those interested in the sometimes bizarre science and occultism practiced at the court of Rudolf II in Prague will find R. J. W. Evans's *Rudolf II and His World: A Study in Intellectual History 1576–1612* (Oxford, 1973) rewarding.

Among general studies relating to Renaissance science must be included the pioneering works by E. A. Burtt, *The Metaphysical Foundations of Modern Physical Science* (rev. ed., London, 1932), and E. W. Strong, *Procedures*

142

and Metaphysics: A Study in the Philosophy of Mathematical-Physical Science in the 16th and 17th Centuries (Berkeley, 1936). Both are still frequently cited. Lynn Thorndike's massive A History of Magic and Experimental Science (8 vols., New York, 1923–1958) emphasizes magic rather than science, but it serves as a rich bibliographic source for anyone interested in the period. George Sarton's no less monumental chronological histories of science never reached the Renaissance, but two short works, Appreciation of Ancient and Medieval Science During the Renaissance (1450–1600) (New York, 1961) and Six Wings: Men of Science in the Renaissance (Bloomington, Ind., 1957), present a considerable amount of material in a characteristically positivistic fashion. W. P. D. Wightman's Science and the Renaissance. An Introduction to the Study of the Emergence of the Sciences in the Sixteenth Century (2 vols., Edinburgh and London, 1962) is a masterful study of both science and medicine that includes a one-volume register of primary sources. The Eighth International Congress of the History of Science at Tours was devoted to Renaissance science and the proceedings were published under the title of the meeting as Sciences de la Renaissance, ed. Jacques Roger (Paris, 1973). Included are papers by well-known authorities covering a broad spectrum of subjects. Richard Foster Jones' Ancients and Moderns: A Study of the Rise of the Scientific Movement in Seventeenth-Century England (1936; 2d ed., St. Louis, 1961) is rather dated in its approach, but it still contains much of interest.

Still useful as a textbook is Herbert Butterfield's The Origins of Modern Science (New York, 1952), which originated in a series of lectures given at Cambridge in the immediate postwar years. More extensive in coverage is The Scientific Renaissance 1450–1630 by Marie Boas (New York, 1962). For those with an interest in the broader ties of Renaissance science with society and other spheres of intellectual activity there is W. P. D. Wightman's Science in a Renaissance Society (London, 1972). An essential reference work for those seeking biographical information is the Dictionary of Scientific Biography, editor-in-chief, Charles C. Gillispie (14 vols., New York, 1970–1976).

Paolo Rossi's Philosophy, Technology and the Arts in the Early Modern Era, trans. Salvator Attansio, ed. Benjamin Nelson (New York, 1970), relates technology to the philosophy and science of the sixteenth and the seventeenth centuries; Bertrand Gille has discussed the Engineers of the Renaissance (Cambridge, Mass., 1966). E. G. R. Taylor has produced a readable history of navigation in The Haven Finding Art (London, 1956); a more detailed study is The Art of Navigation in Tudor and Stuart England (London, 1959) by D. W. Waters. Mathematicians and instrument makers are the subject of E. G. R. Taylor's *The Mathematical Practitioners of Tudor and Stuart England 1485–1714* (Cambridge, 1968). In *Ballistics in the Seventeenth Century: A Study in the Relations of Science and War with Reference Principally to England* (Cambridge, 1952) A. R. Hall has also devoted considerable space to pre-seventeenth-century material.

In addition to Taylor, the student of Renaissance mathematics will wish to read Paul Lawrence Rose's recent *The Italian Renaissance of Mathematics: Studies on Humanists and Mathematicians from Petrarch to Galileo* (Geneva, 1975). The impact of Nicolaus Cusanus, important for his neo-Platonism as well for the influence of his cosmology and mathematics, is best studied through primary sources. *Of Learned Ignorance* has been translated by Fr. Germain Heron (London, 1954) and an interesting anthology including the weight experiments from *The Idiot* has been prepared by John P. Dolan: *Unity and Reform: Selected Writings of Nicholas de Cusa* (South Bend, Ind., 1962). Scientific autobiography from this period is scarce, but an outstanding example exists for the polymath Jerome Cardan in *The Book of My Life,* trans. Jean Stoner (New York, 1930).

A classic treatment of the neo-Platonic influence is Arthur O. Lovejoy's *The Great Chain of Being* (Cambridge, Mass., 1936), but although this is still stimulating reading, the reader interested in magic in the early modern period should also turn to more recent works, including D. P. Walker's *Spiritual and Demonic Magic from Ficino to Campanella* (London, 1958), Charles G. Nauert's *Agrippa and the Crisis of Renaissance Thought* (Urbana, Ill., 1965) and Keith Thomas's *Religion and the Decline of Magic: Studies in Popular Beliefs in Sixteenth- and Seventeenth-Century England* (London, 1971). A collection of essays including much of interest on Hermeticism and alchemy as well as more common scientific and medical themes is *Science, Medicine and Society in the Renaissance,* ed. Allen G. Debus (2 vols., New York, 1972). For primary sources from the literature of natural magic there is J. B. Porta's *Natural Magick* (English translation of 1658; rpt. New York, 1957) and H. C. Agrippa's *Three Books of Occult Philosophy or Magic: Book 1 – Natural Magic,* ed. Willis F. Whitehead (1897; rpt. London, 1971).

John Dee remains a special case of an author whose work belongs to mathematics and astronomy no less than to alchemy, astrology, and spiritualism. The most recent study is that of Peter J. French, whose *John Dee: The World of an Elizabethan Magus* (London, 1972) has much on Dee's mysticism, but does not treat his "real science" adequately. Dee's influential *The Mathematicall Praeface to the Elements of Geometrie of Euclid of Megara* (1570) has been reprinted with an introduction by Allen G. Debus (New

York, 1975), and the secret diaries of his association with the alchemist John Kelly and his attempts to contact the spirit world have been reprinted in *A True and Faithful Relation of What Passed for Many Years Between Dr. John Dee . . . and Some Spirits . . . with a Preface by Meric Casaubon* (London, 1659; rpt. Glasgow, 1974).

There is no satisfactory history of astrology for this period, but the reader will find much of interest relating to English astrological debates in Don Cameron Allen's *The Star-Crossed Renaissance: The Quarrel About Astrology and its Influence in England* (1941; rpt. New York, 1966). In contrast, alchemy has been treated by many authors. For the general reader E. J. Holmyard's Penguin volume, *Alchemy* (Harmondsworth, 1957), will serve as a fascinating primer. Those interested in the connection of this subject with primitive folk beliefs and metalworking technology will wish to go on to Mircea Eliade's *The Forge and the Crucible,* trans. Stephen Corrin (New York, 1962). Allen G. Debus has prepared a short survey in his article on "Alchemy" in the *Dictionary of the History of Ideas,* ed. Philip P. Weiner (4 vols., New York, 1973), I, 27–34. For those who might wish to sample the original texts, Elias Ashmole's compilation is the best and most convenient source in English [*Theatrum Chemicum Britannicum* (1652), rpt. with an introduction by Allen G. Debus (New York and London, 1967)].

The Renaissance chemical technology associated with mining techniques and metallurgy is best located through a series of important translations that have appeared in this century. Georgius Agricola's *De re metallica* was translated by Herbert Clark Hoover (then mining engineer, later President) and his wife, Lou H. Hoover, in 1912 (rpt. New York, 1950). No less important is the series of translations prepared by Cyril Stanley Smith. Here we note only Vannochio Biringuccio's *Pirotechnia,* trans. and ed. by C. S. Smith and M. T. Gnudi (1942; rpt. Cambridge, Mass., 1966) and Lazarus Ercker's *Treatise on Ores and Assaying,* trans. A. G. Sisco and C. S. Smith (Chicago, 1951). All these translations have valuable introductions.

Essential background to the iatrochemical revolt of Paracelsus is found in Owsei Temkin's *Galenism: Rise and Decline of a Medical Philosophy* (Ithaca, N.Y., 1973). The key study of Paracelsus is Walter Pagel's *Paracelsus: An Introduction to Philosophical Medicine in the Era of the Renaissance* (Basel and New York, 1958). This may be supplemented by Paracelsus, *Selected Writings,* ed. Jolande Jacobi, trans. Norbert Guterman (New York, 1951), and Paracelsus, *Volumen Medicinae Paramirum,* trans. with a preface by Kurt F. Leidecker (Baltimore, 1949).

The world view of the Paracelsians is discussed by Allen G. Debus in *The Chemical Philosophy: Paracelsian Science and Medicine in the Sixteenth and Sev-*

enteenth Centuries (2 vols., New York, 1977). Other pertinent studies by Debus include "The Chemical Philosophers: Chemical Medicine from Paracelsus to Van Helmont," History of Science, 12 (1974), 235–59; The English Paracelsians (London, 1965); "Mathematics and Nature in the Chemical Texts of the Renaissance," Ambix, 15 (1968), 1–28, 211; "Motion in the Chemical Texts of the Renaissance," Isis, 64 (1973), 4–17; and "Renaissance Chemistry and the Work of Robert Fludd" in Alchemy and Chemistry in the Seventeenth Century: Papers Read at a Clark Library Seminar, March 12, 1966 (Los Angeles, 1966). Those preferring an early account of the chemical philosophy should seek out Oswald Croll's "Discovering the Great and Deep Mysteries of Nature," which was translated from the admonitory preface of the Basilica Chymica (1609) by H. Pinnell and included in his Philosophy Reformed and Improved in Four Profound Tractates (London, 1657). A reprint with an introduction by Allen G. Debus is planned for the near future. Owen Hannaway has sought the origins of the discipline of chemistry in his study of the conflicting views of Croll and Andreas Libavius in The Chemists and the Word: The Didactic Origins of Chemistry (Baltimore and London, 1975). For an idea of the understanding of geocosmic events by Renaissance chemists, see Frank Dawson Adams' The Birth and Development of the Geological Sciences (1938; rpt. New York, 1954).

The strong influence of chemical methods on traditional plant lore may be seen in the reprint of Hieronymus Brunschwig's Book of Distillation (English trans., c. 1530), introduction by Harold J. Abrahams (New York, 1971). Few of the major herbals have been reprinted in their entirety. An exception is John Parkinson's Paradisi in Sole, Paradisus Terrestris, or a Garden of All Sorts of Pleasant Flowers Which Our English Ayre Will Permit (1629; New York, 1975). The high quality of botanical illustrations in late antiquity may be seen in the magnificent elephant folio edition of the Codex Iulianae picturis illustratus (Dioscorides) (2 vols., Leiden, 1906).

The best survey of the herbal literature remains Agnes Arber's Herbals: Their Origin and Evolution. A Chapter in the History of Botany (Cambridge, 1912), but C. E. Raven's English Naturalists from Neckham to Ray (Cambridge, 1947) is useful. An essential recent study is Karen M. Reed's "Renaissance Humanism and Botany," Annals of Science, 33 (1976), 519–42. Jerry Stannard has many articles relating to medieval and early modern botany. The reader will find particularly useful "The Herbal as a Medical Document," Bulletin of the History of Medicine, 43 (1969), 212–26; "P. A. Mattioli: Sixteenth Century Commentator on Dioscorides," University of Kansas Bibliographical Contributions, 1 (1969), 59–81; and "Medieval Herbals and Their Development," Clio Medica, 9 (1974), 23–33. For a more popular account see Eleanour S. Rohde's *The Old English Herbals* (1922; rpt. New York, 1971).

The influence of the new discoveries on European natural history forms a prominent part of Donald F. Lach's monumental *Asia in the Making of Europe* (to date, 2 vols. in 5 parts, Chicago, 1965–1977). A study of special importance is Alfred W. Crosby's *The Columbian Exchange: Biological and Cultural Consequences of 1492* (Westport, Conn., 1972). C. R. Boxer's *Two Pioneers of Tropical Medicine: Garcia d'Orta and Nicolás Monardes* (London, 1963) is short but illuminating. Both d'Orta's *Colloquies on the Simples and Drugs of India*, trans. with an introduction by Sir Clements Markham (London, 1913), and Monardes's *Joyfull Newes Out of the Newe Found Land* (London, 1925) are available in English.

There are fewer primary sources relating to Renaissance animal lore than there are to herbals. However, Edward Topsell's *Historie of Four-Footed Beastes* (1607) and *Historie of Serpents* (1608) – both of which are based primarily on the work of Conrad Gesner – have been reprinted twice in recent years (New York, 1967; Norwood, N.J., 1973). Selections appear in *The Elizabethan Zoo* (London, 1926). Also useful are E. Callot's *La Renaissance des sciences de la vie au XVIe siècle* (Paris, 1951) and E. J. Cole's *A History of Comparative Anatomy* (London, 1944). Still to be admired for its choice of material if not for its positivistic bias is Charles Singer's *A History of Biology: A General Introduction to the Study of Living Things* (rev. ed., New York, 1950).

The discovery of the circulation of the blood has been discussed by many authors, but the short account most frequently cited is that of Charles Singer, *The Discovery of the Circulation of the Blood* (1922; rpt. London, 1956). A useful survey will be found in Mark Graubard, *Circulation and Respiration: the Evolution of an Idea* (New York and Burlingame, Calif., 1964). A short account that places Harvey in a broader context of medical history is W. P. D. Wightman's *The Emergence of Scientific Medicine* (Edinburgh, 1971).

On more specific topics Ludwig Choulant is the author of the important *History and Bibliography of Anatomical Illustration*, trans. and ed. Mortimer Frank (Chicago, 1920). C. D. O'Malley's *Andreas Vesalius of Brussels 1514–1564* (Berkeley and Los Angeles, 1964) is the standard biography of that author in English, and his *Michael Servetus. A Translation of His Geographical, Medical and Astrological Writings* (Philadelphia, 1953) presents the most interesting texts of that enigmatic author. The most important study of Harvey is Walter Pagel's *William Harvey's Biological Ideas: Selected Aspects and Historical Background* (Basel and New York, 1967), which he has supplemented with a second volume, entitled *New Light on William*

Harvey (Basel, 1976). The reader may also wish to read the books of Kenneth David Keele, *William Harvey: the Man, the Physician, and the Scientist* (London, 1965) and Gweneth Whitteridge, *William Harvey and the Circulation of the Blood* (London and New York, 1971). Robert Fludd's reaction to the work of Harvey has been discussed by Allen G. Debus in two papers: "Robert Fludd and the Circulation of the Blood," *Journal of the History of Medicine and Allied Sciences,* 16 (1961), 374–93; "Harvey and Fludd: The Irrational Factor in the Rational Science of the Seventeenth Century," *Journal of the History of Biology,* 3 (1970), 81–105. There are a number of different translations of Harvey's work in print. Specialists consider none of them to be faultless, but any of them will give the reader some feeling of Harvey's style and accomplishment. The most readily available is the translation of Robert Willis (1847), in the Everyman series: William Harvey, *Circulation of the Blood and Other Writings* (New York and London, 1952). Harvey's debt to Aristotle and Galen may be readily seen in his section on scientific method, which serves as an introduction to the *Anatomical Exercises on the Generation of Animals* in *The Works of William Harvey M.D.,* trans. Robert Willis (London, 1847), pp. 151–67.

Surprisingly little has been published on the influence of Harvey on medicine even though all refer to the fundamental nature of the discovery. Two important exceptions are Audrey B. Davis, *Circulation Physiology and Medical Chemistry in England 1650–1680* (Lawrence, Kan., 1973), and Pedro Laín Entralgo, "La obra de William Harvey y sus consecuencias," in P. Laín Entralgo (ed.), *Historia universal de la medicina* (Barcelona, 1973).

The most arresting single figure in Renaissance surgery is Ambroise Paré. His autobiography is available in *The Apologie and Treatise of Ambroise Paré Containing the Voyages Made unto Divers Places, with Many of his Writings on Surgery,* ed. with an introduction by Geoffrey Keynes (Chicago, 1952). A seventeenth-century translation of his massive works has recently been reprinted as *The Collected Works of Ambroise Paré,* trans. Thomas Johnson (1634) (Pound Ridge, N.Y., 1968).

All authors interested in the Scientific Revolution have had to deal with the new astronomy and its consequences for the physical sciences. Yet among the general accounts none has fully displaced J. L. E. Dreyer's *A History of Planetary Systems from Thales to Kepler* (Cambridge, 1906). Essential among the more recent studies is Alexandre Koyré's *The Astronomical Revolution: Copernicus, Kepler, Borelli,* trans. R. E. W. Maddison (Ithaca, N.Y., 1973). On a more elementary level are the excellent works of Thomas S. Kuhn, *The Copernican Revolution. Planetary Astronomy in the Development of Western Thought* (Cambridge, Mass., 1957), and I. Bernard

Cohen, *The Birth of a New Physics* (Garden City, N.Y., 1960). A pioneering study relating to the acceptance of the Copernican theory in a national setting is Francis R. Johnson's *Astronomical Thought in Renaissance England: A Study of the English Scientific Writings from 1500 to 1645* (Baltimore, 1937).

Angus Armitage's *Copernicus: The Founder of Modern Astronomy* (New York, 1962) provides a useful introduction to his life and work; the *Commentariolus* is available in translations by Edward Rosen (New York, 1939) and Noel M. Swerdlow (Philadelphia, 1973). The Rosen edition includes a translation of the *Narratio prima* by Rheticus as well as the *Commentariolus*. The *De revolutionibus* itself has been most recently translated by A. M. Duncan (London, 1977).

Fundamental for understanding the change from the late medieval to the early modern world view is Alexandre Koyré's *From the Closed World to the Infinite Universe* (New York, 1958). Frances A. Yates has discussed the world view of Bruno in her *Giordano Bruno and the Hermetic Tradition* (Chicago, 1964), a book that has been extremely influential in pointing out the need of considering Hermetic and mystical themes in the rise of modern science. Recent criticisms of her position are to be found in Robert S. Westman, "Magical Reform and Astronomical Reform: The Yates Thesis Reconsidered," in Robert S. Westman and J. E. McGuire, *Hermeticism and the Scientific Revolution* (Los Angeles, 1977), pp. 5–91; Brian Vickers, "Francis Yates and the Writing of History," *Journal of Modern History, 51* (1979), 287–316; and Allen G. Debus, "The 'Pseudo-Sciences' and the History of Science," *The University of Chicago Library Society Journal, 3* (1978), 3–20. For those interested in the spectrum of magnetic views presented by William Gilbert there is D.H.D. Roller's *The De Magnete of William Gilbert* (Amsterdam, 1959). This may be supplemented with Gilbert's *De magnete*, trans. P. Fleury Mottelay (1893; rpt. New York, 1958).

Like his more general work on planetary theory, J. L. E. Dreyer's *Life of Tycho Brahe: A Picture of Scientific Life and Thought in the Sixteenth Century* (Edinburgh, 1890) remains a classic. For Kepler see the recent survey of the literature in E. J. Aiton's "Johannes Kelper in the Light of Recent Research," *History of Sciences, 14* (1976), 77–100. The standard biography is that of Max Caspar, trans. and ed. by C. Doris Hellman (London and New York, 1959), but Angus Armitage's *John Kepler* (London, 1966) is also useful. For those particularly interested in the interplay of mysticism and science in the work of Kepler there is Arthur Koestler's popular *The Watershed: A Biography of Johannes Kepler* (Garden City, N.Y., 1960).

The literature on Galileo is vast and much of it is of high quality.

Perhaps the most influential work on this author in this century has been Alexandre Koyré's *Etudes Galiléennes* (3 vols., Paris, 1939), as yet untranslated, but the reader will find that the series of papers in *Galileo: Man of Science*, ed. Ernan McMullin (New York, 1967) presents the spectrum of recent attitudes toward this author. Outstanding among recent studies are William R. Shea's *Galileo's Intellectual Revolution: Middle Period, 1610–1632* (New York, 1972) and Maurice Clavelin's *The Natural Philosophy of Galileo: Essay on the Origins and Formation of Classical Mechanics*, trans. A. J. Pomerans (Cambridge, Mass., 1974). Stillman Drake's *Galileo Studies: Personality, Tradition, and Revolution* (Ann Arbor, Mich., 1970) presents a collection of essays by an acknowledged leader in this field.

Many of the most important works by Galileo are available in translation. The *Discoveries and Opinions of Galileo,* trans, and with an introduction and notes by Stillman Drake (Garden City, N.Y., 1957), includes the *Sidereus nuncius* (1610), the *Letters on Sunspots* (1613), the *Letter to the Grand Duchess Christina* (1615) and excerpts from *The Assayer* (1623). *The Dialogue Concerning the Two Chief World Systems* (1632) is available in translations by Stillman Drake (Berkeley and Los Angeles, 1953) and Giorgio de Santillana (Chicago, 1953). Similarly the *Mathematical Discourses and Demonstrations Concerning Two New Sciences* (1638) is available in two translations, that of Henry Crew and Alfonso de Salvio (1914; rpt. New York, 1954) and that of Stillman Drake (Madison, Wis., 1974).

The most frequently cited recent study of Francis Bacon is Benjamin Farrington's *The Philosophy of Francis Bacon* (Liverpool, 1964), but this account is to be supplemented with Paolo Rossi's examination of the Hermetic influence seen in the less commonly read tracts of Bacon in his *Francis Bacon: From Magic to Science*, trans. Sacha Rabinovitch (Chicago, 1968). Bacon's link with Paracelsus has been examined by Graham Rees in "Francis Bacon's Semi-Paracelsian Cosmology," *Ambix,* 22 (1975), 81–101. Nearly all of Bacon's philosophical and scientific works were translated in the collected edition by Spedding and Ellis (1857), and these translations were gathered into one volume by John M. Robertson: *The Philosophical Works of Francis Bacon* (London and New York, 1905).

Descartes has always been of special interest to philosophers, but to date relatively few historians of science have discussed his work in detail. One of the few works of this sort is J. F. Scott's *The Scientific Work of René Descartes (1596–1650)* (London, n.d.). Also of interest is Henri Gouhier's *Les premières pensées de Descartes: contribution à l'histoire de l'anti-renaissance* (Paris, 1958). The views of Scott and Gouhier are to be contrasted with William R. Shea's "Descartes and the Rosicrucians," *Annali dell' Instituto e Musec di Storia della Scienza di Firenze,* 4 (1979), 29–47. Descartes' *Discourse on*

Method is available in any number of acceptable translations. One of the most convenient is that of F. E. Sutcliffe, published by Penguin (Baltimore, 1968).

For those most concerned with the triumph of the mechanical philosophy E. J. Dijksterhuis' *The Mechanization of the World Picture*, trans. C. Dikshoorne (Oxford, 1961), is required reading. More difficult is René Dugas' valuable *Mechanics in the Seventeenth Century*, trans. J. R. Maddox (Neuchatel, 1958). R. S. Westfall has prepared a short account of the triumph of the mechanical philosophy in the major scientific fields in his *The Construction of Modern Science: Mechanisms and Mechanics* (New York, 1971). Robert Lenoble's *Mersenne ou la naissance du mécanisme* (1943; rpt. Paris, 1971) goes into considerable detail on Mersenne's debates, his correspondence, and his intellectual development. Of the many works on the revival of atomism in the seventeenth century the reader is likely to find Andrew G. Van Melsen's *From Atomos to Atom* (New York, 1960) and Robert H. Kargon's *Atomism in England from Hariot to Newton* (Oxford, 1966) of most interest. Gassendi's attempt to carry out the suggested experiments of Galileo is described by Allen G. Debus in "Pierre Gassendi and His 'Scientific Expedition' of 1640," *Archives internationales d'histoire des sciences,* 16 (1963), 129–42.

The scientific utopias of the early seventeenth century are surveyed by Nell Eurich in *Science in Utopia: A Mighty Design* (Cambridge, Mass., 1967). The Rosicrucian literature is discussed by Paul Arnold, *Histoire des Rose-Croix et les origines de la Franc-Maçonnerie* (Paris, 1945) and more recently by Frances A. Yates in *The Rosicrucian Enlightenment* (London and Boston, 1972). All works on the Rosicrucians are to be used with caution. A persistent theme in the account of Yates is the ascription of most key features of the Scientific Revolution to a mystical origin. The seventeenth-century English translation of the basic Rosicrucian texts was reprinted at Margate in 1923 as *The Fame and Confession of the Fraternity of R:C: Commonly of the Rosie Cross,* and Felix Emil Held translated Johann Valentin Andreae's extremely interesting *Christianopolis: An Ideal State of the Seventeenth Century* (New York, 1916).

The only major work by Fludd in English is his *Mosaicall Philosophy* (London, 1659), but J. B. Craven's *Doctor Robert Fludd (Robertus de Fluctibus) The English Rosicrucian. Life and Writings* (Kirkwall, 1902), though old, is still useful. Allen G. Debus' "Robert Fludd and the Use of Gilbert's *De magnete* in the Weapon-Slave Controversy," *Journal of the History of Medicine and Allied Sciences,* 19 (1964), 389–417, and "The Sun in the Universe of Robert Fludd," in *Le soleil à la Renaissance — sciences et mythes, Travaux de l'Institut pour l'étude de la Renaissance et de l'Humanisme,* 2 (1964), pp.

257–78, deal with significant aspects of Fludd's thought. Wolfgang Pauli has a perceptive study of the Kepler–Fludd exchange in "The influence of Archetypal Ideas on the Scientific Theories of Kepler," in C. G. Jung and W. Pauli, *The Interpretation of Nature and the Psyche*, trans. Priscilla Silz (New York, 1955), pp. 147–240. For van Helmont see Allen G. Debus, "The Chemical Debates of the Seventeenth Century: The Reaction to Robert Fludd and Jean Baptiste van Helmont," in M. L. Righini Bonelli and William R. Shea, eds., *Reason, Experiment and Mysticism in the Scientific Revolution* (New York, 1975), pp. 18–47, 291–8.

The texts of John Webster, Seth Ward, John Wilkins, and Thomas Hall relating to educational reform in 1654 have been reprinted by Allen G. Debus in *Science and Education in the Seventeenth Century, The Webster-Ward Debate* (London and New York, 1970), and the agricultural plans of the chemists are discussed by Debus in "Palissy, Plat and English Agricultural Chemistry in the 16th and 17th Centuries," *Archives internationales d'histoire des sciences,* 21 (1968), 67–88. Glauber's chemical and economic plans for Germany are discussed in Debus, *The Chemical Philosophy,* 2, pp. 425–41. The relationship of science to society in mid-seventeenth century England has been well covered by Charles Webster in *The Great·Instauration: Science, Medicine and Reform 1626–1660* (New York, 1976).

And finally, for those having a special interest in Isaac Newton's alchemy and Hermeticism as evidence of the continued influence of characteristically Renaissance themes, there is J. E. McGuire and P. M. Rattansi's "Newton and the 'Pipes of Pan,' " *Notes and Records of the Royal Society of London,* 21 (1966), 108–43, and Allen G. Debus, "Van Helmont and Newton's Third Law" in *Paracelsus, Werk und Wirkung. Festgabe für Kurt Goldammer zum 60. Geburtstag,* ed. Sepp Domandl, *Salzburger Beiträge zur Paracelsusforschung,* 13 (Vienna, 1975), pp. 45–52. B. J. T. Dobbs has discussed *The Foundations of Newton's Alchemy, or "The Hunting of the Greene Lyon"* (Cambridge, 1975), which should be read along with the substantial review by Karin Figala, "Newton as Alchemist," *History of Science,* 15 (1977), 102–37.

Sources of Quotations

page

1   Francis Bacon, *The Works of Francis Bacon*, ed. James Spedding, Robert Leslie Ellis, and Douglas Dennon Heath (7 vols., new ed., London: Longmans, 1870; original ed. 1857), 4, p. 114.

3   Ramus quotation from Frank Pierrepont Graves, *Peter Ramus and the Educational Reformation of the Sixteenth Century* (New York: Macmillan, 1912), pp. 23–4.

10   Galileo Galilei, *Dialogues Concerning Two New Sciences,* trans. Henry Crew and Alfonso de Salvio (New York: Dover, 1954), p. 1.

17–18   Bonus of Ferrara, *The New Pearl of Great Price,* trans. A. E. Waite (London: James Elliott, 1894; reprinted London: Vincent Stuart, 1963), p. 138.

20   Paracelsus, *Selected Writings,* trans. Norbert Guterman, ed. Jolande Jacobi (New York: Bollingen Series XXVIII, Pantheon Books, 1951), pp. 79–80.

47   d'Orta quotations from C. R. Boxer, *Two Pioneers of Tropical Medicine: Garcia d'Orta and Nicolás Monardes* (London: The Hispanic and Luso-Brazilian Councils, 1963), p. 14.

51   Bock and Zaluziansky quotations from Agnes Arber, *Herbals, Their Origin and Evolution, A Chapter in the History of Botany* (Cambridge: Cambridge University Press, 1912), pp. 136, 151.

60, 63   Vesalius quotations from Charles Singer, *A History of Biology* (New York: Henry Schuman, 1950, revised ed.), p. 103.

63   Vesalius quotation on his Galenist critics from C. D.

O'Malley, *Andreas Vesalius of Brussels 1514–1564* (Berkeley and Los Angeles: University of California Press, 1964), p. 222.

67, 69   William Harvey, *The Circulation of the Blood and Other Writings* (London: J. M. Dent/New York: E. P. Dutton, Everyman Edition, 1952), pp. 56, 57, 85.

78, 79   Ficino and Melanchthon quotations from Thomas S. Kuhn, *The Copernican Revolution: Planetary Astronomy in the Development of Western Thought* (Cambridge, Mass.: Harvard University Press, 1957), pp. 129, 191.

82   Copernicus quotation from Alexandre Koyré, *From the Closed World to the Infinite Universe* (New York: Harper Torchbook, 1957), p. 33.

107   Descartes quotation from A. Rupert Hall, *From Galileo to Newton 1630–1720* (New York: Harper & Row, 1963), p. 193.

109, 110   Galileo Galilei, *Dialoges Concerning Two New Sciences,* trans. Henry Crew and Alfonso de Salvio (New York: Dover, 1954), pp. 166, 160, 162.

120, 121   Andreae quotations from J. V. Andreae, *Christianopolis. An Ideal State of the Seventeenth Century,* trans. F. E. Held (New York: Oxford University Press, 1916), pp. 137–8, 187, 196–7.

Index

Printed in the United States
67035LVS00005B/190-285

9 780521 293280